Hamlyn all-colour paperbacks

Biology
John M. Hard

illustrated by David Pratt

Hamlyn Paperbacks

FOREWORD

The aim of this book is to provide a simple introduction to some of the more important aspects of biology. It should be emphasized, however, that the subject is such a large one that many topics have had to be omitted or given only scant mention. Such aspects, for example genetics, evolution and natural selection, are covered very adequately by other books in this series.

Wherever possible the subjects under discussion have been described from the biological point of view, which has avoided separating plants and animals into two completely false and arbitrary categories.

J.H.

Published by Hamlyn Paperbacks
The Hamlyn Publishing Group Limited
Astronaut House, Feltham, Middlesex, England

ISBN 0 600 31313 1

Phototypeset by Filmtype Services Limited, Scarborough
Colour separations by Colour Workshop Limited, Hertford
Printed in Spain by Mateu Cromo, Madrid

CONTENTS

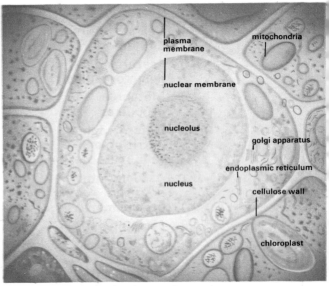

Plant cell as shown by the electronmicrograph.

CELL STRUCTURE AND FUNCTION

With relatively few exceptions, organisms are built from cells. Plant cells are often easier to study as they tend to be larger and have a fixed shape. The cell wall of most plant cells is built from a number of layers of material, the basic building substance being cellulose. Despite the large amount of material incorporated into the plant cell wall it remains permeable to the movement of water and gases. An exception to such permeability is the corky bark cell (see page 59) where the thick layer of impermeable material results in the death of the cell contents, the vast mass of dead cells providing the plant with an inert protective jacket.

Under the cellulose cell wall lies the living plasma membrane, a very thin sheet of material which, in section, is thought to consist of a complex triple layered structure. The three layers being a 'fat sandwich' in which the 'bread' is protein and the 'filling' fat. The plasma membrane is the outer limiting layer of animal cells, and, as in plants, it shows great powers of active selective absorption and excretion.

Within the cell lies a number of different structures. The **nucleus**

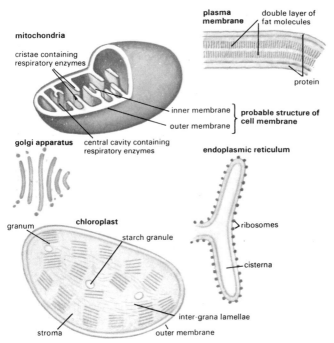

Detail of some cell organelles.

is the control centre of the cell, and by means of chemical messages it regulates cell metabolism. The nucleic acids which form the vast bulk of the nucleus are organized into chromosomes, the structures involved in the physical movements of cell division. The nucleus is bounded by the nuclear membrane built along the same lines as the plasma membrane. Perforations in the nuclear membrane are thought to permit entry and exit of the chemical messages initiated by the chromosomes. The **endoplasmic reticulum** consists of a branching network of fine tubes and cavities throughout the cell. It provides a large surface area for the attachment of *ribosomes*, the minute structures responsible for protein assembly. The reticulum is continuous with the nuclear membrane. *Golgi apparatus* is the name given to a specialized part of the endoplasmic reticulum consisting of flattened cavities parallel to the nuclear membrane. Its function appears to be one of packaging and exporting materials from the cell. **Mitochondria** are the centres of respiratory activity.

They have a double skin structure, the inner one being characteristically infolded. **Chloroplasts** are pigment-containing structures found in plant cells. Their function is to assist in photosynthesis. Other cell structures include *lysosomes* – packets of enzymes, and *centrioles* – a bundle of nine rods which assist in cell division.

FEATURES OF ALL LIVING ORGANISMS

There are a number of phenomena exhibited by all living organisms.

Respiration This is the means by which plants and animals obtain energy, derived from chemical activity within their cells. In addition to the important cell chemistry there may also be bodily movements associated with the exchange of gases.

Nutrition The source of the energy to be liberated in respiration is the materials taken into the organism. In animals this is an obvious feeding activity, in plants the less obvious process of photosynthesis using direct sunlight energy is used.

Growth As an organism ages, particularly in the early stages of its life, it will increase in dimension and complexity. Animals typically show some limit to growth, i.e. they reach 'adult' size. Plants, however, may well add to their bulk throughout life.

Movement In most cases animals can move their entire body; plants show a more restricted movement of only parts of their structure, e.g. petals. A number of the microscopic algae are capable of free movement of their body.

Irritability Living organisms respond to their environment in a number of ways, e.g. loud sounds cause obvious responses in some animals; unilateral light causes responses in plants. As a result of responses to regular environmental changes, organisms often exhibit rhythmic or cyclic behaviour patterns, e.g. the opening and closing of flowers, migration of birds.

Excretion The various physical and chemical activities of an organism result in waste materials such as carbon dioxide and ammonia, and these must be eliminated to prevent poisoning of the organism.

Reproduction It is an obvious necessity, for continuation of the species, that an organism should have a means of reproduction.

Death Eventually all organisms come to the end of their life and the various processes mentioned above, cease. The constituents of the body have not finished their role, however, as the processes of decomposition result in a recycling of the various minerals.

Structure of the cellulose cell wall.

Differences between plants and animals

There is an increasing realization of the futility of attempting a sharp separation of plants from animals, particularly at the microscopic level. There are, however, a few features worthy of note that appear to be peculiar to one group or the other in the vast majority of cases.

Most plants possess chlorophyll, and by the process of photosynthesis are capable of manufacturing their food from simple inorganic materials (see page 8). Animals do not possess chlorophyll and have to feed upon complex organic foods.

Plant cells tend to have a fixed shape once the major period of cell differentiation has finished. This rigidity is due to the building material cellulose being incorporated into the cell walls.

Animal cells are limited by a living, pliable plasma membrane which is capable of changes of shape throughout its functional life.

Although all organisms exhibit sensitivity, the speed of response varies greatly. Animals respond promptly whereas plants are far slower and their response may take hours or even days. It is a feature of plant responses that their slow movements are limited to certain regions.

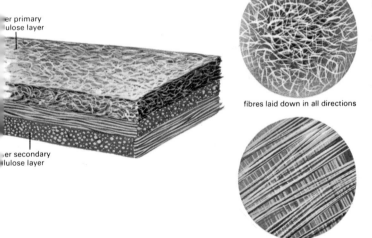

er primary
ulose layer

er secondary
lulose layer

fibres laid down in all directions

fibres laid down parallel

METABOLIC PROCESSES
Photosynthesis

This is the phenomenon, taking place in green plants, which is the fundamental light energy-trapping process.

Briefly, chlorophyll 'traps' light and uses its energy to synthesize carbohydrates (sugars) from such simple materials as water and carbon dioxide. During this process the water molecule is split, yielding oxygen as a by-product. It is this oxygen that is utilized as the essential respiratory gas by most organisms. The summary equation for photosynthesis is as follows:

$$6 CO_2 + 6 H_2O \xrightarrow[\text{chlorophyll}]{\text{light energy}} C_6 H_{12} O_6 + 6 O_2.$$

carbon dioxide   water   glucose   oxygen

This is a gross oversimplification of a complex series of cycles and reactions, however. As chlorophyll is essential, obviously photosynthesis only occurs in the green parts of a plant. From the diagram it can be seen that the major photosynthetic region of a leaf is the palisade layer located in the upper part of the leaf. The lower half is more concerned with gaseous exchange within the leaf. The leaf is flattened to give the maximum surface of chloroplasts exposed to sunlight with the minimum distance for gaseous diffusion.

The pores on the lower surface – the stomata – are sites of entry and exit of gases, the evaporation of water being the penalty for free flow of carbon dioxide. The carbon dioxide is used up in the chloroplasts and creates a concentration gradient leading from the high concentration of the atmosphere via the stomata, which occur mainly on the lower half of the leaf, into cells containing the active chloroplasts. Oxygen diffuses from actively photosynthesizing cells via the same route leading to the atmosphere.

Water, the other essential material, is brought to the leaf in the xylem vessels, the water transport route of the vascular bundles.

Not all the light falling on the chlorophyll is of value in photosynthesis and thus not all the different wavelengths of sunlight are absorbed. Only those associated with photosynthesis are absorbed to any great extent, i.e. in the red and blue wavelengths. This absorption pattern is seen clearly if a beam of white light is passed through an extract of chlorophyll and then viewed through a spectroscope. This instrument passes the incoming light through a lens and prism system thus separating the different wavelengths.

(*Top*) Leaf structure. (*Bottom*) Absorption spectrum of chlorophyll; the black areas indicate those wavelengths of light which disappear after being passed through a chlorophyll extract.

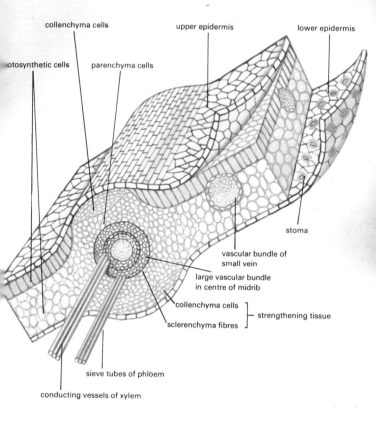

collenchyma cells

otosynthetic cells

parenchyma cells

upper epidermis

lower epidermis

stoma

vascular bundle of
small vein

large vascular bundle
in centre of midrib

collenchyma cells

sclerenchyma fibres

strengthening tissue

sieve tubes of phloem

conducting vessels of xylem

450   500   550   600   650   700   750 nm

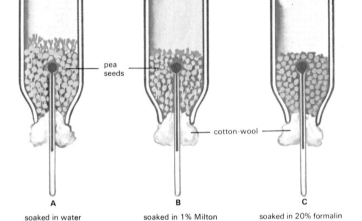

pea
seeds

cotton-wool

A
soaked in water

B
soaked in 1% Milton

C
soaked in 20% formalin

(*Above*) Pea seeds were soaked in different solutions and then kept in damp blotting paper for twenty-four hours to start germination, before being placed in flasks. (*Below*) Results of heat production.

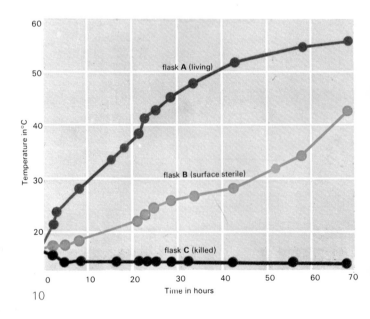

flask **A** (living)

flask **B** (surface sterile)

flask **C** (killed)

Temperature in °C

Time in hours

Respiration

Respiration is the complex process resulting in energy being made available to an organism. There are two basic forms of respiration: **aerobic**, in the presence of, and requiring oxygen, e.g. most organisms; and **anaerobic**, in the absence of oxygen, e.g. yeasts and a few internal parasites.

Both these forms of respiration yield energy, but the aerobic form is many times more energy productive and hence is used by any active organism. The low energy yield of anaerobic respiration is in fact derived during the early stages of aerobic respiration.

In aerobic respiration there are two main stages: **external respiration**, the simple gaseous exchanges in and out of the organism (breathing movements); and **internal respiration**, the biochemistry that occurs within a respiring cell resulting in energy being made available from the material oxidized.

The chemistry of respiration is complex, but in essence consists of a series of oxidations resulting in a simple sugar (glucose) being broken down to its constituent units of water and carbon dioxide. During this process oxygen is required and energy is made available to the organism. A simple summary equation of aerobic respiration is as follows:

$$C_6H_{12}O_6 + 6\ O_2 \longrightarrow 6\ CO_2 + 6\ H_2O + \text{energy}.$$

glucose   oxygen   carbon dioxide   water

Although at first sight this may appear to be the reverse of photosynthesis, it is certainly not so, the chemical pathways involved in these two processes being different. Anaerobic respiration involves the incomplete breakdown of glucose, the equation reading:

$$C_6H_{12}O_6 \longrightarrow 2\ C_2H_5OH + 2\ CO_2 + \text{little energy}.$$

glucose   ethyl alcohol   carbon dioxide

Not only is the energy production low, but the ethyl alcohol by-product is toxic and will kill most living tissue. Obviously any organism capable of living on a very low energy demand, whilst in the midst of ethyl alcohol, is following a highly specialized life. One group of organisms that regularly respire anaerobically are the yeasts – in this case these fungi are of considerable value to man who uses the ethyl alcohol in brewing and the carbon dioxide in baking.

Despite the disadvantages of anaerobic respiration, active tissue will be forced to respire in this way if deprived of adequate oxygen. Examples could be active muscle fibres which have temporarily consumed all their oxygen, or even the central tissue of a large apple where the rate of oxygen diffusion inwards is inadequate due to sheer bulk of tissue.

In the experiment illustrated, some of the respiratory energy is liberated as heat, an inefficient event, but useful as a demonstration of respiratory production.

Finally, it should be remembered that although all living organisms respire continually, the oxygen replenished by photosynthesis is derived only in sunlight. Obviously photosynthesis must occur much faster if the oxygen:carbon dioxide balance is to remain constant.

Osmosis, diffusion and roots

Atoms and molecules of any material are constantly moving – a phenomenon known as *Brownian motion*. If two different concentrations of a solution are placed in a container, this Brownian motion results in an even distribution of the molecules, i.e. a final solution with a concentration lying somewhere between the two original concentrations. It is customary to refer to this movement of molecules from a region of high concentration to a region of lower concentration, as moving along a concentration gradient.

If two different concentrations of sugar solution are separated by a membrane, the molecules will bump into the membrane, due to their Brownian motion. Any small holes in the membrane will permit the flow of molecules through the membrane. The only limitation to free flow of molecules will be the size of the molecules in relationship to the hole. Thus in the case of the sugar solutions, the water molecules may be able to move freely in both directions, but the sugar molecules may well be too large. In such an example, the membrane would be described as *semipermeable* as it permitted the free movement of only some molecules.

It is still not fully understood how osmosis works, but the explanation above is a convenient model even if not a completely accurate picture of the true mechanism.

The term diffusion is applied to the process whereby molecules move along a concentration gradient. In the specialized case of limited diffusion, due to the presence of a semipermeable membrane, the term osmosis is used. Osmosis may be defined as the passage of molecules of a solvent from the region of their high concentration, through a semipermeable membrane, to the region having a higher concentration of solute material. The solute, in the case quoted, was sugar; the solvent being water.

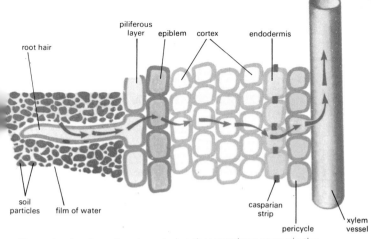

The path taken by soil water entering the vascular system via the root.

In a typical plant cell the cellulose cell wall is permeable to water and most other solutions, thus providing no restriction to molecules. The cytoplasm acts as a semipermeable membrane. The characteristic cell vacuole contains a watery sap having a number of materials dissolved in it. Therefore, sap has a higher concentration of materials than water and is described as having a higher osmotic pressure than that of water.

The root system of a plant has two major functions: **anchorage**, to withstand the tugging caused by wind action on the aerial portions of the plant; and **uptake of water and minerals**, achieved by the mixture of passive water entry plus active uptake of certain chemicals.

The removal of water from the outer root hairs causes its osmotic pressure to rise again, thus enhancing further entry of soil water. Water passes into the root tissue, as illustrated, and enters the xylem vessels in which it rises to the aerial regions. This flow of water into xylem vessels causing its level to rise within the vascular tissue is termed *root pressure*.

The entry of water and salts is obviously enhanced by the large surface area of a thin root, and this surface is further enlarged by the development of fine unicellular outgrowths – root hairs – near the root endings.

The cell sap of the root hairs has a higher osmotic pressure than soil water, cytoplasm acts as a semipermeable membrane, and thus water enters the root hairs by osmosis. The sap of the root hairs is now diluted, and of a lower osmotic pressure than that of the adjacent inner cells. This situation induces an osmotic gradient with water flow from the outer layer (root hairs) into the inner cells.

HETEROTROPHIC NUTRITION AND FOOD MATERIALS

Green plants are capable of making their own complex materials by the process of photosynthesis and the uptake of other compounds, e.g. nitrates, from the soil. A few bacteria can also synthesize complex organic materials from carbon dioxide and water but use chemical energy from cell metabolism, and not light energy as in photosynthesis. Such bacteria are said to exhibit *chemosynthesis*.

It will be noted that both photosynthesis and chemosynthesis are ways of building complex materials from simple compounds. Any organism that can synthesize in this fashion is called *autotrophic*. Organisms such as fungi lack chlorophyll and therefore cannot photosynthesize; they are also incapable of energy-trapping by chemosynthesis. Their method of obtaining energy is by the intake of complex materials and breaking these down to simpler compounds, thus liberating energy. Such a breakdown method of obtaining energy is termed *heterotrophic*.

Types of food and their function

Before going further into the question of energy supply and demand, and its relationship to food, the major constituents of man's diet must be listed. These are: **carbohydrates**, the major energy-supplying foods; **proteins**, body building materials; **fats**, energy-rich food stores; **vitamins**, vital components of cell chemistry; **mineral salts**, raw materials of all metabolism; **water**, an obvious and essential compound required for internal cell pressure, the basic medium of cell chemistry and major constituent of blood and hence the transport system; and **roughage**, the mass of food, mainly composed of plant cellulose, that gives the solid bulk to food and thus enables the gut muscles to grip it and move it along by peristalsis.

As has been mentioned on pages 9 and 10, the amount of energy liberated depends greatly upon the method of respiration used to oxidize the food material. There are, however, two other features to be considered, namely the varying amount of energy available in different foods, and the energy requirements of different organisms under different situations.

Carbohydrate sources

The three categories of common food materials, carbohydrate, protein and fat differ in energy yield:

1 gramme of carbohydrate yields 17·2 kilojoules of heat
1   ,,   of protein   ,,   22·2   ,,   ,, ,,
1   ,,   of fat   ,,   38·5   ,,   ,, ,,

From these figures it is obvious that fat is a very rich energy source, and an ideal storage material, as a small quantity can obviously supply large amounts of energy. Although protein supplies more heat energy than carbohydrate per gramme burnt, there are problems in breaking down protein as an energy source. This is due to the production of toxic waste products – a highly undesirable feature. The waste products of fat and carbohydrate respiration are water and carbon dioxide, both of which are easily and safely dealt with.

Man's energy requirements vary enormously. There is a basic minimum required (basal metabolic rate) simply to keep the body alive and functional. Any active effort will cause the energy demand to rise as can be seen in the chart below.

At rest (asleep)	4·2 kilojoules per minute	
Sitting up (woman)	5·0	,, ,, ,,
Sitting up (man)	5·2	,, ,, ,,
Light work (man)	14·5	,, ,, ,,
,, ,, (woman)	17·5	,, ,, ,,
Heavy work (man)	43·0–52·0 ,,	,, ,,

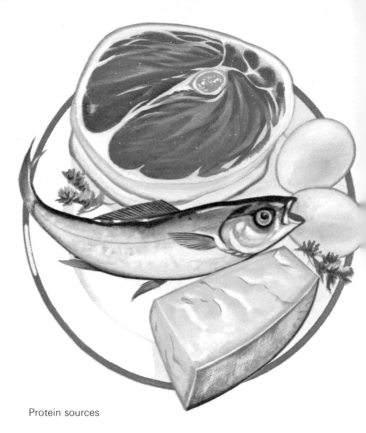

Protein sources

The structure and functions of food materials

Carbohydrates

The illustration on page 15 shows a few common sources of carbo-hydrates, the following lines give some indication of the chemical nature of such materials. Carbohydrates contain three elements – carbon, hydrogen and oxygen; the elements hydrogen and oxygen are present in units of water (H_2O). There are three main cate-gories of carbohydrates.

MONOSACCHARIDES (single sugars), e.g. glucose. These are the important sugars actually burnt in respiration. All monosaccharides have the chemical formula $C_6H_{12}O_6$.

DISACCHARIDES (commercial granulated sugar), e.g. sucrose. These are the sugars widely known to the public as sweeteners of tea, cakes, etc. They are built up from two units of monosaccharides, and all have the chemical formula $C_{12}H_{22}O_{11}$. Disaccharides are built up as follows:

$$C_6H_{12}O_6 + C_6H_{12}O_6 \longrightarrow C_{12}H_{22}O_{11} + H_2O.$$

   glucose   glucose   maltose   water

POLYSACCHARIDES, e.g. starch, cellulose. These are complex molecules built from many smaller carbohydrate units, e.g. starch has the chemical formula $(C_6H_{10}O_5)n$ where n may be several hundred. Starch is a storage material formed in plants, as a result of photosynthesis. Cellulose is a building material, already mentioned in the account of plant characteristics.

Proteins

The constituent elements of proteins are carbon, hydrogen, oxygen and nitrogen, plus usually sulphur and very often phosphorus. There may be other elements present in some proteins, e.g. magnesium. These elements are combined to form the basic 'building bricks' of any protein – those units known as amino acids. Amino acids link together to form proteins in much the same way as monosaccharides join to form polysaccharides. When two amino acids join they form a unit called a dipeptide; further amino acid units join forming a larger complex, a polypeptide. Polypeptides are long chains of amino acids, and proteins are formed when adjacent polypeptide chains cross-link.

Proteins are used for the synthesis and repair of cells, and hence are essential for healthy life and growth. They also provide the raw materials for the synthesis of enzymes – the catalysts of cell chemistry. Unfortunately proteins cannot be stored in the body, and therefore a steady supply is required.

Fats

Like carbohydrates, fats contain the elements carbon, hydrogen and oxygen, but in different proportions from the sugars. As has been seen on page 15, fats are very rich sources of energy and this is their main value in the life of most animals. Animals store fat under the skin, around organs, such as the kidneys, and occasionally in specialized fat bodies as seen in insects and amphibians. A thick layer of fat has considerable insulation value against environmental cold, as well as being a potential energy source, and is found in many cold water aquatic forms, e.g. whales and walruses.

Vitamin sources

Vitamins

The chemistry of vitamins is complex, and suffice it to say that their presence is essential for healthy cell metabolism in any organism. Plants are capable of synthesizing their vitamins, whereas animals require them in their diet. Two exceptions to this last statement are the vitamins manufactured by gut bacteria, from which the 'host' benefits, and vitamin D, which can be synthesized in the body and then activated by sunlight or ultra-violet light.

Tests for different food materials

In the previous pages a wide range of materials has been introduced, many of which can be tested for in a mixed food source by a series of simple chemical tests.

GLUCOSE If glucose solution is heated to boiling point with Benedict's reagent the colour changes from the blue of Benedict's, through a range of greens and yellows to the brick-red precipitate of cuprous oxide.

SUCROSE This fails to give a positive Benedict's test until the double sugar is split into its constituents by boiling with dilute hydrochloric acid. The boiled solution is neutralized with sodium bicarbonate, after which a positive Benedict's test is obtained.

PROTEIN Solid protein, such as eggwhite, should be boiled with Millon's reagent, when a pink-red mass results, indicating a positive test.

FATS A few drops of Sudan black should be added to a fat, e.g. castor oil, and then the tube shaken, forming an emulsion. On leaving for a while the fat drops coalesce at the surface taking the black dye with them.

Minerals

Mineral salts are required in very small quantities, but their absence is dramatic and serious in its consequences. This can best be illustrated by some common examples.

CALCIUM, PHOSPHORUS and MAGNESIUM are all required by animals for healthy teeth and bone formation. Magnesium is also used in chlorophyll formation in plants. Poor seed germination results from phosphate deficiency.

IRON is a constituent of haemoglobin, the red pigment in blood corpuscles. A deficiency results in a form of anaemia. Iron is also required for the formation of chlorophyll.

SODIUM and POTASSIUM are essential for efficient nerve functioning and the general cell metabolism in animals. A deficiency in plants results in problems for photosynthesis.

Certain elements are required by both plants and animals for the same functions, e.g. NITROGEN and SULPHUR are needed for protein synthesis and phosphorus for protein and nucleic acid synthesis.

Although the total mass of all of the above chemicals may remain relatively constant in an environment, their availability will vary as the different materials are used, discarded, synthesized or 'locked up' in different compounds.

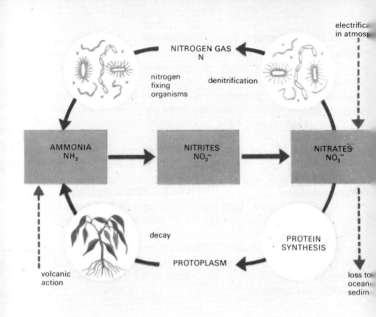

The nitrogen cycle

The cycle of minerals in nature

As stated earlier, most elements are present in most environments, but they are not always readily available.

Minerals and man

Great emphasis is placed these days on the need to produce the best possible crops in the greatest possible quantity from farmland. Such production is possible only if people realize that the crops remove a great deal from the soil, and what is taken must be replaced if future crops and the health of the soil are to be guaranteed.

Addition of humus Humus is a term used to cover any dead organic material. Numerous organisms, for instance worms, bacteria and fungi, break such humus down further, and their activities release materials from the humic debris. In addition to the chemical value of humus, it can improve the texture of soils greatly, e.g. in wet, heavy soils, humus breaks the solid wet clumps into soil crumbs so improving aeration and drainage. In dry, sandy soils, humus improves not only the chemical content but also the water content of the soil by retaining some of the water as it drains through the sand.

Addition of artificial fertilizers This practice has the value of permitting the farmer to add only those chemicals in which his soil is deficient. The disadvantage of artificial fertilizers, however, is their short-term nature as they do not improve the physical structure of the soil.

Rotation of crops A very ancient and useful method of aiding the soil is by growing different crops each year on the same field. The reason is that different crops require different nutrients in varying amounts, e.g. wheat takes more nutrient than other cereals. In any cycle of crop rotation, a leguminous crop like clover should be included, as it improves the content of soil nutrients, especially nitrates (see illustration of nitrogen cycle). Such leguminous crops may be ploughed in.

Addition of lime Lime's action on soil is two-fold. Firstly, it improves crumb structure in soil and this improves aeration and drainage. Secondly, it renders the soil more alkaline which encourages more micro-organisms, and also permits plants to take up minerals that are otherwise insoluble in acid soils.

The carbon cycle

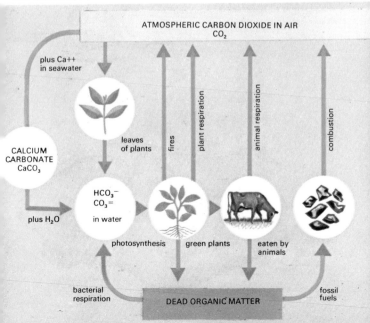

ATMOSPHERIC CARBON DIOXIDE IN AIR CO_2

plus Ca^{++} in seawater

CALCIUM CARBONATE $CaCO_3$

leaves of plants

fires

plant respiration

animal respiration

combustion

HCO_3^- $CO_3^=$ in water

plus H_2O

photosynthesis

green plants

eaten by animals

bacterial respiration

DEAD ORGANIC MATTER

fossil fuels

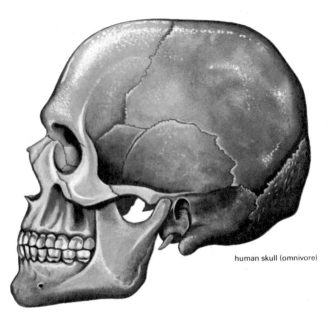

human skull (omnivore)

Variations in the type and
distribution of teeth are a
result of different diets.

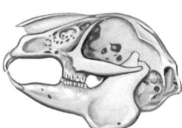

rabbit skull (herbivore)

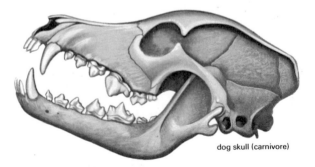

dog skull (carnivore)

Feeding and dentition

Feeding in animals is an essentially active process involving a number of stages.

Capture of prey This is very variable and ranges from the simple engulfing of prey by part of the predator, e.g. *Amoeba* using pseudopodia, through the trapping of prey using tentacles, e.g. *Hydra*, *Octopus*, to the calculated pursuit and killing of prey, as seen in the big cats.

Ingestion The entry of prey, or parts torn from the prey, is usually via a specialized opening – the mouth. There are animals which can take food into the body at any point, e.g. *Amoeba*, but this is by no means common, even in unicellular animals (protozoa). Many of the protozoa have a specialized region of the body surface for entry of food – the cytostome, and once ingested, the food follows a set pathway through the cell during its digestion. This passage through the cell, or cyclosis, is seen very well in *Paramecium* (see page 46).

Digestion Most animals have some form of gut.

Egestion The voiding of indigestible remains.

Types of teeth

The mammals show a very clear specialization of their dentition depending upon their particular diet.

INCISORS These are chisel-shaped teeth at the front of the mouth used for cutting and nipping.

CANINES Pointed, fang-like teeth used for killing prey and tearing flesh.

PREMOLARS and MOLARS The 'cheek' teeth situated at the sides and back of the jaws variously surfaced and used for crushing and grinding.

Modifications of dentition to suit diet

The herbivores (plant eaters) have large, sharp incisors for cutting off vegetation, and occasionally, some may be missing leaving a pad of tough skin against which the opposite incisors cut, e.g. the cow. Canines are missing leaving a characteristic gap (diastema). The cheek teeth are large, surfaced with ridges of enamel and dentine and used for the reduction of vegetation to a pulp.

The carnivores (flesh eaters) have small, conical incisors which are rarely of such value to the animal as in the herbivores. The canines are very pronounced and used for killing and tearing at prey. The cheek teeth are reduced in number, sharply pointed and covered in a layer of enamel. Some cheek teeth can be used for shearing when

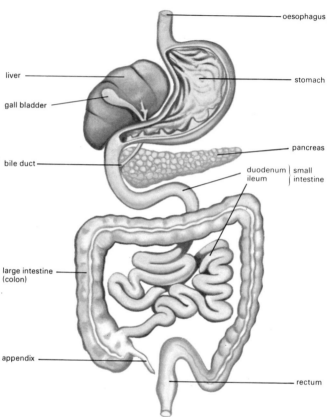

oesophagus
liver
gall bladder
stomach
bile duct
pancreas
duodenum | small
ileum | intestine
large intestine (colon)
appendix
rectum

(*Above*) Human digestive system. (*Below*) Stomach of ruminant.

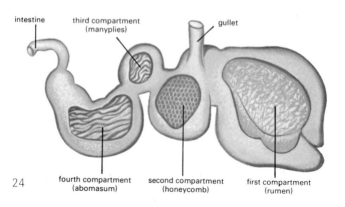

intestine
third compartment (manyplies)
gullet
fourth compartment (abomasum)
second compartment (honeycomb)
first compartment (rumen)

upper and lower jaw teeth slide past each other like the blades of scissors.

Omnivores (animals which eat both flesh and vegetation) will clearly require a full complement of the different types of teeth. Their incisors are chisel-shaped, the canines present although smaller than carnivores', and the cheek teeth smaller than those of herbivores and usually enamel covered. Whatever the type of teeth possessed by an animal it should be remembered that they all contribute to the same ends, namely, reduction of food to smaller pieces for swallowing, and by making food lumps smaller, increasing the surface area of food exposed to enzyme attack.

Digestion and the gut

Once the physical stage of digestion – the attack by the teeth – is finished, the reduction of food becomes chemical.

With increasing complexity of life there is a parallel increase in complexity of body organs, not least of which is the gut. The simplest 'gut' is probably the food vacuole of the protozoa. This is closely parallelled by the 'simple' gastric chamber of the coelenterates. In both protozoa and coelenterates there is a region in which all enzymes function. Clearly there must be organization in the production of enzymes for any one set of conditions would prove unsatisfactory for many enzymes. In *Amoeba* the vacuole content varies from acid at first, during which protein digestion takes place, to alkaline for fat and carbohydrate breakdown. Coelenterates are often more haphazard in their acidity regulation, and it is doubtful if the optimum conditions for any one set of enzymes ever prevail.

As animals became increasingly complex, there slowly evolved a specialized tubular gut. Furthermore, it is found that the gut is *regional*, i.e. divided into sections, each region having a major function.

The first region of the mammalian gut is the buccal cavity which is divided horizontally into two regions – the food canal and the respiratory canal – by the hard palate. This enables a mammal to chew and breathe simultaneously. In the buccal cavity food is chewed and moistened with saliva. The saliva acts as a lubricant and also contains an enzyme, ptyalin, which breaks down starch into maltose, a disaccharide sugar. After chewing, the food is swallowed and passes down the oesophagus into the stomach. The action of muscles contracting and squeezing food down the gut is called peristalsis.

The stomach of man exemplifies many of the characteristics

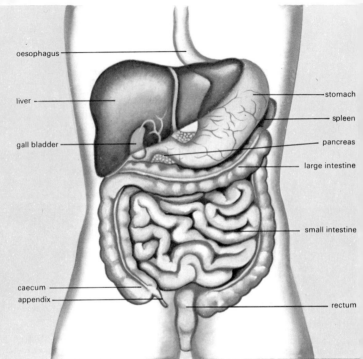

oesophagus

liver

gall bladder

stomach

spleen

pancreas

large intestine

small intestine

caecum

appendix

rectum

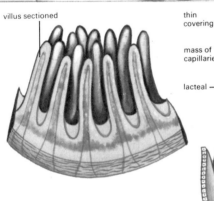

villus sectioned

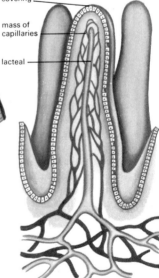

thin covering

mass of capillaries

lacteal

(*Top*) Human digestive system *in situ*. (*Above*) Part of the lining of the small intestine showing villi. (*Right*) Villus sectioned.

common to mammals and therefore its secretions and activities will be described.

Secretions of the stomach

Hydrochloric acid	sterilizes food, starts protein breakdown and provides correct medium for pepsin.
Pepsin (enzyme)	converts protein into peptones.
Mucus	lubricates and protects stomach lining from acid and pepsin.
Rennin (enzyme)	in young mammals, an aid to milk digestion.

Vigorous stomach contractions churn the food to and fro until all is thoroughly mixed with the secretions.

The stomach of ruminants (cattle, sheep, deer and goats) differs from other mammals in having four chambers. The habit of chewing the cud is also seen. Grass first bitten off passes to the rumen and then to the reticulum. From here it returns by antiperistalsis to the mouth for further chewing, to reduce its bulk further. Food is reduced to a semi-solid pulp and swallowed a second time. It now passes to the third chamber, the manyplies, which has a number of folds used for straining the food, before being passed to the fourth chamber, the true stomach, or abomasum.

In the first chamber, the rumen, bacterial digestion of cellulose occurs; this is invaluable as the ruminant cannot produce cellulose enzymes, and hence it relies on these beneficial bacteria. The abomasum is the only chamber to produce gastric secretions.

From the mammal's stomach food passes to the small intestine which is divided into two regions, the duodenum and the ileum. The duodenum is a relatively short length of gut but into it pass two sets of important secretions. From the gall bladder, via the bile duct, comes bile. This yellow, alkaline fluid reduces the acidity of the food and also emulsifies fats, i.e. breaks them down into smaller droplets. Pancreatic juices, secreted by the pancreas, flow into the duodenum. These juices contain a number of enzymes which are concerned with the breakdown of the three main classes of food.

A little food is absorbed in the duodenum, but the bulk of the products of digestion are absorbed in the ileum which produces its enzymes as part of the intestinal juices. Uptake of food in the small intestine is aided by the large surface area, provided by the numerous finger-like outgrowths, the *villi*. The secretion of mucus and alkaline fluids is from Brunner's glands, whilst cells lining the crypt of Lieberkühn produce enzymes, which are listed on page 28.

	ENZYME	SUBSTRATE	PRODUCTS OF DIGESTION
Pancreatic Juices	Amylase	Starch	Maltose
	Lipase	Fat	Fatty acids +Glycerol
	Trypsin	Protein	Amino acids
Intestinal Juices	Lipase	Fat	Fatty acids +Glycerol
	Maltase	Maltose	Glucose
	Sucrase	Sucrose	Glucose +Fructose
	Peptidase	Peptides	Amino acids

After passing through the small intestine, most of the useful food has been digested and absorbed. The indigestible remains, plus a quantity of mucus and various digestive fluids, pass on to the large intestine.

In many animals, the first region, the caecum, is extremely short and food passes to the next region, the colon. The colon is an adaptation for terrestrial life, its function being the absorption of water. The digestive juices are produced in a liquid form, and clearly no land animal can afford to waste such a large volume of water. The action of the colon leaves the indigestible waste as a nearly solid mass, the faeces. The faeces pass to the rectum where they are stored until voided, via the anus.

In non-ruminant herbivores, i.e. vegetation eaters that do not chew the cud before passing it to their stomachs, there is a development of the caecum. This is the first region of the large intestine and develops as a cul-de-sac at the junction of the ileum and colon. Its function is to provide a site for breakdown of cellulose, for in this type of herbivore, it will be remembered, no cellulose digestion has yet occurred. The caecum is rich in bacteria which digest the cellulose, thus releasing cell contents plus the derivatives of cellulose digestion. The material emerging from the caecum is rich in food materials and, in the case of the rabbit, will pass through the gut again when the rabbit eats its mucoid pellets. This re-eating permits the valuable food materials to pass a second time via the small intestine, where absorption can take place.

The gut of an unknown animal can indicate a great deal concerning its way of life and certainly its diet. The overall length and proportions of parts will tell whether it is carnivorous or herbivorous. A long gut with a caecum, or a long gut with a chambered stomach indicates the problem of cellulose digestion – a herbivore. The presence of a long colon could well mean a terrestrial animal.

Comparison of digestive systems of human (*top*) and rabbit (*bottom*).

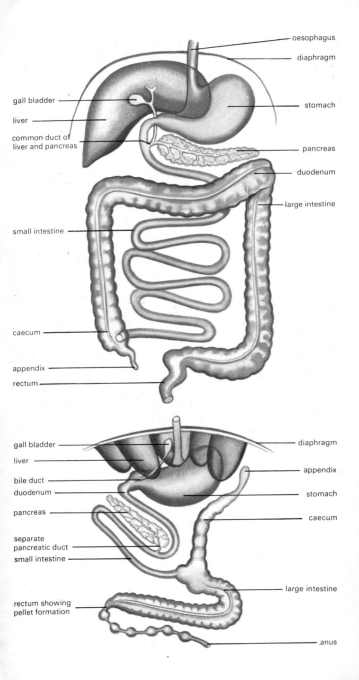

oesophagus

diaphragm

gall bladder

liver

common duct of
liver and pancreas

stomach

pancreas

duodenum

large intestine

small intestine

caecum

appendix

rectum

gall bladder

liver

bile duct

duodenum

pancreas

separate
pancreatic duct

small intestine

rectum showing
pellet formation

diaphragm

appendix

stomach

caecum

large intestine

anus

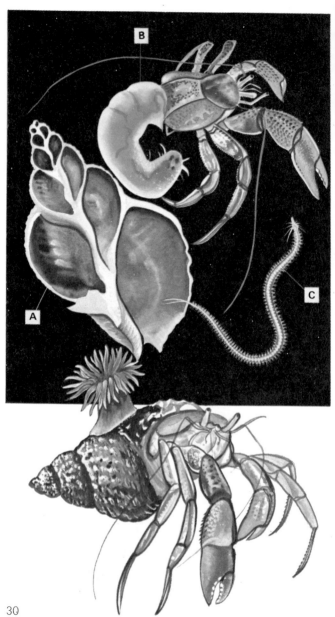

ASSOCIATIONS BETWEEN ORGANISMS

During the course of evolution many organisms have ceased to live a completely free and independent existence and instead are now found living in some form of relationship with one or more organisms. The degree of association varies, from organisms that are capable of living a completely independent life with the potential to form an association, to a mutually dependent association that is essential to both partners. These associations have been given names and are often described as static situations. It should be emphasized, however, that one form of association will often seem to blend into another, living organisms being extremely reluctant to fit neatly into arbitrary classifications.

Commensalism

This state is defined as being an association between two organisms where one benefits but not at the expense of the other. This phenomenon can be an extremely loose relationship, or it can, apparently, be very close to a state where *both* participants could gain, i.e. it would be a mutually beneficial association and therefore moved out of the classification of commensalism.

As an example of commensalism, there is the hermit crab and its 'hitchhikers'. The crab gains no value from the association of many of the other organisms living on its shell home. The residents on the shell, however, gain from any scraps of food the crab drops, and also by being taken into regions that might otherwise be beyond their capabilities. A good example of a crab commensal is the colonial hydroid *Hydractinia*, which fixes itself to rocks and shells but is incapable of colonizing loose mud and sand. The crabs can move over loose substrate and thus *Hydractinia* may capture its own food in a region otherwise inaccessible to it.

Animals can also form an association with plants. For example, a tube-dwelling worm, *Spirorbis*, forms small, spiral white tubes on the common seaweed, the serrate wrack. The larval worm responds to the surface of the wrack, settles down and produces the start of the calcareous tube. The worm benefits by having a suitable home site, the alga remains unaffected.

A commensal association that could possibly be thought of as damaging to the 'host' is found in the hermit crab, where it shares its shell with a bristleworm. The worm emerges when the crab feeds

(*Top*) Sectioned shell of hermit crab (A); hermit crab removed from shell (B); true commensal worm of hermit crab (C). (*Bottom*) Hermit crab with commensal anemone.

and eats some of the food. Could removal of food by the worm be considered harmful?

Another commensal relationship showing a trend this time towards a mutually beneficial association can be found in some of the vertebrate associations. The remora, a sucker fish, is found attached to the skin of sharks and gains protection and transport from its formidable host. The shark may gain nothing but remoras have been known to eat skin parasites of the shark – a case of mutual gain.

Development of other associations from commensalism

The examples of the bristleworm and the remora show that commensalism may develop in two opposite ways. Further predation by the bristle worm might prove harmful to the crab; whilst the establishment of some cleaning arrangement by the remora could yield healthy benefits to the shark and guaranteed protection and food for the remora.

Symbiosis

This term can be confusing as different authorities treat it in different ways. Its literal meaning of 'living together' can cover parasites, commensals and also a mutually beneficial association. To overcome this confusion the term *mutualism* is used here to cover the mutually beneficial relationships.

Mutualism can range from the fleeting, albeit essential, association of two fish, to the intimate, complex and permanent relationship seen in some plants, e.g. lichens. Firstly, the temporary association of a 'cleaner' fish or shrimp with its partner will be considered. In many parts of the world there are animals which gain their food by actively removing growths, parasites and organic debris from other larger organisms. These useful scavengers are cleaning symbionts, an excellent example of mutualism. It has been shown that large fish, e.g. wrasse, deprived of their cleaner associates quickly become diseased. Furthermore the cleaner animals are extensively modified for their way of life – snouts being pointed, obvious body coloration, and the exhibition of a specialized behaviour pattern.

Lichens inhabit and colonize some of the most inhospitable areas of the world, bare rock faces, mountains and polar regions. They are obviously extremely hardy plants, yet they are composed of two types of organism – a fungus and an alga – neither of which

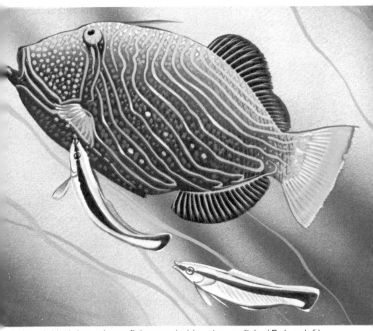

(*Above*) Undulate trigger fish attended by cleaner fish. (*Below left*)
Lichen growing on tree. (*Below right*) Detail of lichen structure as
seen in section.

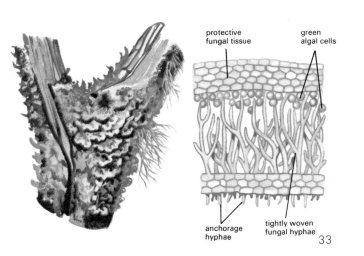

protective
fungal tissue

green
algal cells

anchorage
hyphae

tightly woven
fungal hyphae

could survive in such unpleasant regions alone. The fungus supplies the protective body, and the alga provides growth factors such as glucose, vitamins, alcohols and amino acids.

Parasitism

This relationship between host and associate has been hinted at already and can be defined as the association between organisms whereby only one, the parasite, benefits, the partner suffering some definite harm. There is often some loss of free life on the part of the parasite, and the association between host and parasite is frequently of long duration when judged against the life span of the organisms involved.

Parasites exhibit four features that collectively identify them as such. Firstly, they live in or on a host, and do it harm. The depth to which they penetrate the host varies, as indeed does the damage. Fleas, leeches and lice live on the surface and cause superficial injury. Athlete's foot is a skin disease caused by a fungus living in the surface layers of the foot. The parasite of sleeping sickness is found in the host's blood wriggling between blood corpuscles. Secondly, parasites show some simplification of body structures when compared with free-living relatives. *Sacculina* (a relative of the crab), shows loss of limbs and is reduced to a mass of reproductive tissue within the abdomen of its crustacean host. Dodder, a plant parasite, lacks leaves, roots and chlorophyll. Thirdly, although all

Many of the tapeworm's segments are full of eggs.

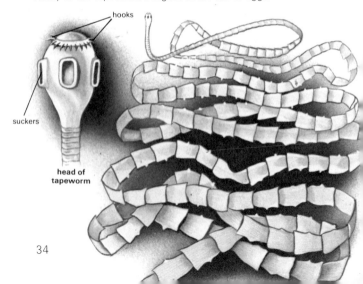

hooks

suckers

head of
tapeworm

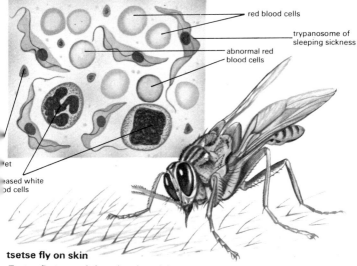

stained specimen of blood from rat infected with sleeping sickness

red blood cells

trypanosome of sleeping sickness

abnormal red blood cells

...et

...ased white ...od cells

tsetse fly on skin

Tsetse fly transmitting sleeping sickness.

organisms show adaptations to their way of life, in the case of parasites they are often associated with a complex physiological response, e.g. the ability to survive in regions almost devoid of available oxygen, such as adult liver flukes, or the hooks and suckers of adult tapeworm. Lastly, parasites exhibit a complex and efficient reproduction, usually associated in some way with the physiology of the host, e.g. rabbit fleas are stimulated by the level of sex hormone in their host.

Many authorities consider that the most damaging and traumatic parasitic associations are probably relatively recent relationships – the participants not yet having had time to 'settle down' to the parasitic way of life. Obviously it is of no value to the parasite to seriously damage, or even kill, its source of food and life.

Parasites have a long history of association with man, and such finds as Egyptian mummies have shown evidence of parasitic infections that were obviously present in people many thousands of years ago. It should be remembered that often the mummies were the remains of wealthy people who presumably lived a full and healthy life by the standards of that time, and hence one can imagine the state of people living in dirty and impoverished conditions.

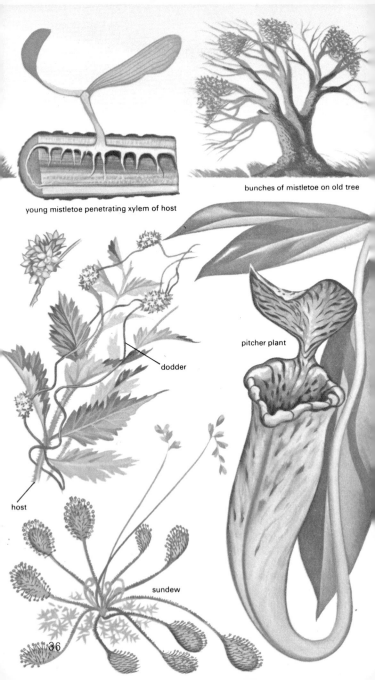

young mistletoe penetrating xylem of host

bunches of mistletoe on old tree

dodder

pitcher plant

host

sundew

36

Heterotrophic nutrition in plants

It has already been stressed that green plants are autotrophic, i.e. synthesize their own food from simple compounds. A few green plants, however, together with the fungi and some bacteria are either incapable of photosynthesis, or need to supplement their diet with additional materials. This last case is exemplified by plants growing in regions deficient in nitrates, e.g. acid bogs.

Parasitic plants

There is a range of plants showing all stages of dependence upon parasitism, from those autotrophs capable of occasional parasitism as a supplement; hemiparasites which always show a mixture of parts; to obligate parasites which rely totally upon parasitism for food.

The yellow rattle is a small plant which 'taps' the root system of grasses. It has green autotrophic leaves and functional roots, and thus the parasitic supply from grasses comes as a supplement.

Mistletoe, found on oak and apple trees, is a mixture, the aerial parts photosynthesize and the roots are modified to act as invasive suckers within host tissue.

Flowering plants lacking chlorophyll are forced to rely on parasitism. The fleshy toothwort is found attached to the roots of hazel and wych elm. The only aerial portion of the toothwort is the flower stem bearing the muddy purple flowers. Another parasitic flowering plant is dodder, a parasite of clover, nettles and several other common plants. Dodder lacks roots, relying upon a twining stem which attaches itself to the host by feeding organs (haustoria).

Saprophytes

These are plants living on dead and decaying organic material, and a good example would be the fungi. Fungi lack chlorophyll and are either parasitic or saprophytic in their nutrition. Some saprophytic fungi form an association with plant roots which is mutually beneficial.

Carnivorous plants

These plants produce a variety of structures which trap animals, usually insects, and then either actively digest the prey, or rely on the decay of the prey. Whatever method is employed, the products diffuse into the carnivorous plant which thus benefits.

Heterotrophic nutrition in plants.

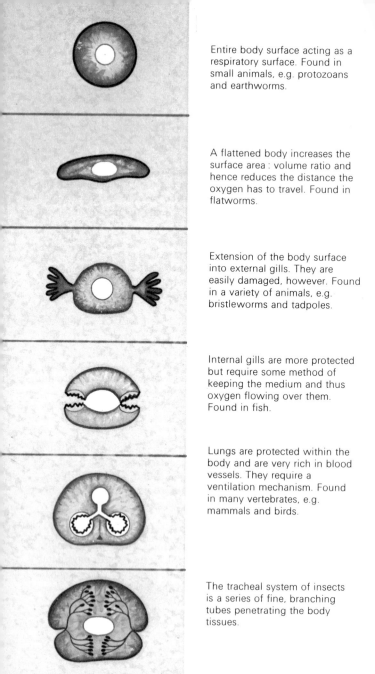

Entire body surface acting as a respiratory surface. Found in small animals, e.g. protozoans and earthworms.

A flattened body increases the surface area : volume ratio and hence reduces the distance the oxygen has to travel. Found in flatworms.

Extension of the body surface into external gills. They are easily damaged, however. Found in a variety of animals, e.g. bristleworms and tadpoles.

Internal gills are more protected but require some method of keeping the medium and thus oxygen flowing over them. Found in fish.

Lungs are protected within the body and are very rich in blood vessels. They require a ventilation mechanism. Found in many vertebrates, e.g. mammals and birds.

The tracheal system of insects is a series of fine, branching tubes penetrating the body tissues.

PRINCIPLES OF RESPIRATION

On page 11 the topic of respiration was briefly covered, the chemistry being described. On the following pages various mechanisms and structures concerned with the external aspects of respiration will be described.

Since the oxygen will have to pass from the atmosphere to the interior of cells, diffusion will be involved at least once. In many cases there will be at least two phases of diffusion: from air into the body via the respiratory surface; and later into cell cytoplasm, often after transport between respiratory surface and respiring cell.

As the respiratory surface is the first major barrier the atmospheric oxygen meets, it is relevant to consider the three criteria to which any surface must conform if it is to act as an efficient respiratory surface.

It must be *thin* as diffusion can only be effective over short distances. It must be *moist* for gases to pass into solution for movement through the membrane. It must be *vascular* to carry gases to and from the surface.

This final point introduces the problem of surface area: volume ratio. A small organism, e.g. *Amoeba*, has a relatively large surface area: volume ratio, whereas a large animal such as a whale has a relatively small surface area: volume ratio. Obviously if only part of the body surface can be used for gaseous exchange, then the large volume becomes even more of a burden. This necessitates some method of ensuring that oxygen at the respiratory surface is immediately taken up and transported, leaving the surface free for further uptake. This is usually a problem solved by some transporting medium which takes the oxygen, either in solution or held by specialized carrier pigments to other parts of the body. Some aquatic animals and a very few terrestrial forms still use entire body surfaces, but with increasing complexity of form, allied to a more active behaviour pattern, it becomes both undesirable and inefficient to use merely the surface. The illustration opposite shows very simply the range of solutions exhibited.

Another problem is the method by which oxygen is carried by the transporting medium; simply in solution means a small amount is carried when compared to the volume of transporting fluid. Respiratory pigments, such as haemoglobin and haemocyanin, are complex chemicals which take up oxygen in regions of high oxygen concentration, carry them, and unload oxygen wherever oxygen is deficient. Such pigments carry far more oxygen, per unit volume, than simply in solution.

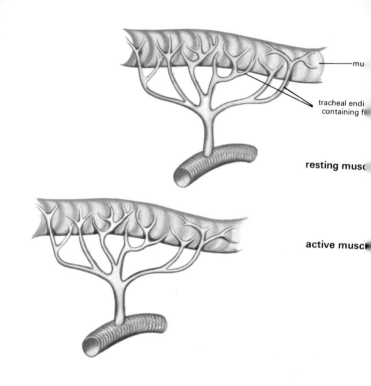

mu

tracheal endi
containing f

resting musc

active musc

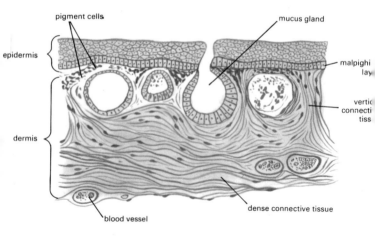

pigment cells

mucus gland

epidermis

malpighi
lay

vertic
connecti
tiss

dermis

blood vessel

dense connective tissue

40

As has already been stated, whether an animal can rely simply on diffusion of gases through its surface depends upon the relationship between its surface area and its volume. If the surface area is large in relationship to its volume, then surface diffusion may well suffice. Such a method demands moist surroundings and hence is found, generally, only in aquatic animals or at least organisms in very damp places.

This last point indicates a failing of surface respiration, namely its limiting nature when it comes to colonization of any terrestrial habitat. This is seen very clearly in the case of the amphibia generally and the frogs in particular. Frogs are limited to damp areas due to the nature of their skin (see illustration opposite). As can be seen, the outer layer, the epidermis, is very thin for quick diffusion. In addition there is a film of watery slime, or mucus, covering the entire surface. The oxygen diffuses, in solution, through mucus and epidermis to the numerous blood capillaries which fulfil the third criterion of a respiratory surface – the need for a transport system. Whilst oxygen combines with the haemoglobin in the red corpuscles, carbon dioxide diffuses from the plasma to the atmosphere in a similar manner.

The amount of oxygen obtained is enough to keep the frog alive as long as it is not active. Should the frog become very active, then the lungs are used to make up the oxgen deficit.

A very similar surface respiration is found in the earthworm: here the epidermis is one cell layer deep and the mucus comes from secretory cells rather than glands. Again there are numerous fine blood capillaries in the epidermal region.

Insects have been very successful land colonists, in no small part due to their efficient and impermeable shell or exoskeleton. Obviously a waterproof covering means surface respiration is impossible, so specialized tubes are found as ingrowths into the insect. These tubes, or tracheoles, lead from very small holes in the exoskeleton, called spiracles, into the centre of the animal's tissues. Movement of gases along the tracheoles by diffusion is very slow and ventilation of the air tubes may occur by bellows-like movements of the abdomen, e.g. in wasps. The fine ends of the tracheoles are fluid-filled when the insect rests, but this can be absorbed when active, permitting gases to flow directly to the active tissues.

(*Top*) Part of the tracheal system of an insect. (*Bottom*) Vertical section through frog skin.

Gills

The gill is an extremely common and variable respiratory organ found throughout the animal world. All gills are basically an area of folded membrane giving the maximum surface area for the minimum volume. They occupy two types of site: projecting from the body (external gills), or sunken into the body tissues (internal gills).

External gills

The very nature of a respiratory surface and the function of a respiratory organ mean that the external gill is a particularly delicate and vulnerable organ. Its presence will normally limit the distribution and habitat of its user to an aquatic existence.

External gills are found in a range of animals, e.g. bristle worms such as the lugworm, early stages of tadpoles and some aquatic insects such as mayfly larvae. Some external gills are partially enclosed by part of the animal's body. This is shown very clearly in the crustacea, e.g. crayfish. In the crustacea, the shell wraps over the delicate gills which are simply outgrowths of the exoskeleton originating from the bases of the legs and the nearby body surface.

Internal gills

Fish gills are derived from the pharynx, and consist of a gill support to which are attached numerous filaments. These filaments have an enormous surface area, due to their extremely folded nature. Furthermore, gills are extremely vascular, being very well supplied with blood capillaries.

Water enters through the mouth, passes through the buccal cavity into the gill pouches and over the filament surfaces, and either directly (dogfish), or indirectly (bony fish), to the exterior.

As water flows over the filament it is very close to blood flowing in the filament capillaries. The flow of water and blood are in opposite directions – a counterflow system. The advantage of this counterflow system is that the blood is constantly coming close to increasingly oxygenated water. Obviously at the point when the blood has reached the base of the filament it is reasonably highly oxygenated. Clearly it would be inefficient if it were now to be in close proximity to water having a low oxygen concentration, as oxygen diffuses along a concentration gradient and would flow from blood to water. The counterflow system ensures that the most highly oxygenated blood is close to even more highly oxygenated water, and thus the concentration gradient of oxygen from water to blood is maintained. The converse is true for carbon dioxide.

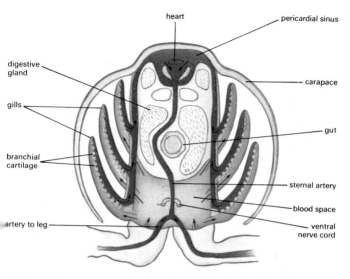

(*Above*) Transverse section through crayfish showing blood flow.
(*Below*) Detail of gill structure.

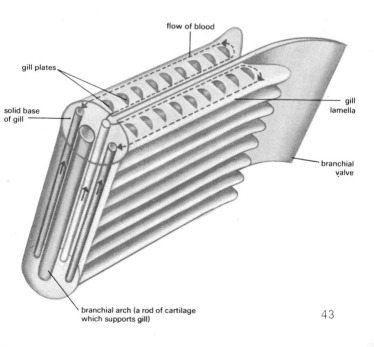

43

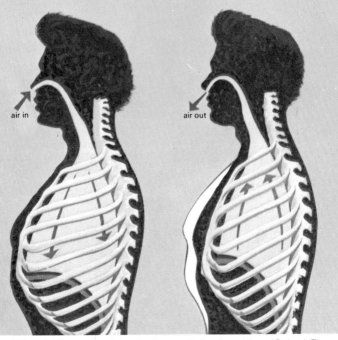

INSPIRATION

EXPIRATION

air in

air out

(*Above*) Movement of the diaphragm during breathing. (*Below*) The lungs and air sacs of a bird.

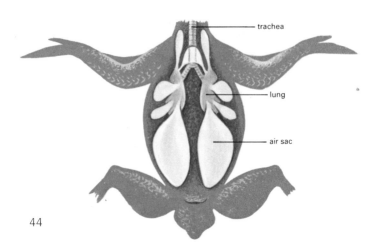

trachea

lung

air sac

Respiration in man

The respiratory system of man consists of a wind pipe which divides into two bronchi, each of which divides many times. At the end of the ultimate divisions are masses of tiny air sacs (alveoli). The lungs consist of a spongy mass of alveoli and tubes leading to the air sacs.

The lungs are sealed in the thorax by pleural membranes, the thorax being separated from the abdomen by the muscular diaphragm. The walls of the thorax are formed by the rib cage and the intercostal muscles.

During inspiration, air enters the lungs when the diaphragm muscles contract, flattening the diaphragm and compressing the abdominal organs. All this increases the thoracic volume from top to bottom. At the same time, the external intercostal muscles contract swinging the ribs up and out thus increasing the thoracic volume from side to side, and from front to back. The overall increase in thoracic volume results in a drop in pressure within and thus air pressure, being greater, forces air into the lungs.

During expiration, the reverse of rib and diaphragm movements, aided by movements of the abdominal contents, an increase in the thoracic air pressure occurs, forcing air out of the lungs.

Movement of gases between the alveolar cavity and blood, via the alveolar and capillary walls, takes place by diffusion, in solution along a concentration gradient, i.e. both carbon dioxide and oxygen move from a region of high concentration to a region of low concentration. The inspired air is richer in oxygen than blood and hence the oxygen molecules pass from the alveolus through the various walls to the red blood cells for transport. Carbon dioxide diffuses from the blood plasma to the cavity of the alveolus for expiration.

Respiration in birds

The thorax of a bird differs from that of a mammal in two ways. Firstly, there is no diaphragm separating the thorax from the abdomen, and secondly, there are paired air sacs branching from the lungs. The air sacs occupy a great deal of room, and estimates indicate that approximately eighty percent of inspired air goes into the paired abdominal air sacs alone. The air sacs are not respiratory surfaces – this function is performed by the lungs alone.

At rest, birds respire using intercostal muscles. In flight, the massive wing muscles exert pressure on the body cavity causing pressure changes within, similar to a diaphragm.

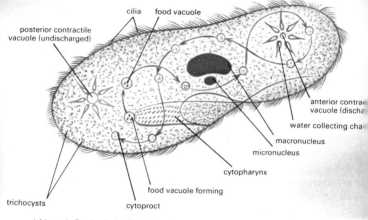

Labels on figure:
posterior contractile vacuole (undischarged)
cilia
food vacuole
anterior contractile vacuole (discharged)
water collecting chamber
macronucleus
micronucleus
cytopharynx
food vacuole forming
cytoproct
trichocysts

(*Above*) Cyclosis in *Paramecium*.
(*Below*) Xylem in higher plants and blood vessels in animals.

MOVEMENT OF MATERIALS IN ANIMALS AND PLANTS

The complex chemistry of an animal's body requires a constant supply of raw materials, the transport of valuable manufactured products, and the removal of waste residues.

The increasing size and complexity of animals has necessitated the development of an active transport system as simple passage of materials by diffusion is impossible over the distance involved.

It should be stressed that the final exchanges of materials between the blood vascular system and the site of metabolism is by diffusion, since the distances involved will be very short – often only the thickness of the cell membrane.

Cyclosis

Movement within a cell is by diffusion; this is shown very clearly in *Amoeba* when the products of digestion within the food vacuole pass by diffusion to all parts of the cell. Such a system is adequate in protistans (unicellular organisms) owing to the very short distances involved. Many protistans, and also individual cells of multicellular organisms, exhibit a form of intracellular transport, however. This consists of a flow of cytoplasm within the cell thereby carrying the organelles and particles along. Rates at which particles move vary and thus mixing of cytoplasmic contents must result.

Cyclosis is seen very clearly in the freshwater alga *Nitella* and midrib cells of Canadian pondweed. A refinement of cyclosis is exhibited by the freshwater protozoan *Paramecium*. In *Paramecium* food enters the body via a specific point, the cytostome, and leaves at another region, the cytoproct. The route between cytostome and cytoproct is relatively fixed and is a form of cyclosis.

Multicellular forms

Most multicellular organisms have some form of piped transport with a variety of means to ensure propulsion along the vessels.

In arthropods – the group including insects, crustaceans and spiders – the blood is unpiped for much of the time. Circulation involves a route of: heart → short vessels → open body spaces → heart. Flow via open spaces of the body is directional due to a series of internal partitions.

The internal organs of the arthropod body are thus bathed in blood, carrying food, whilst receiving oxygen directly from the air via a piped supply such as the tracheole system of insects.

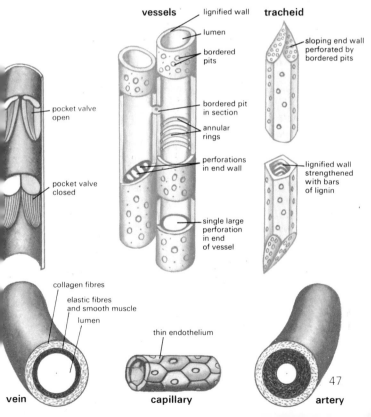

pocket valve open

pocket valve closed

vessels
lignified wall
lumen
bordered pits
bordered pit in section
annular rings
perforations in end wall
single large perforation in end of vessel

tracheid
sloping end wall perforated by bordered pits
lignified wall strengthened with bars of lignin

collagen fibres
elastic fibres and smooth muscle
lumen

thin endothelium

vein

capillary

artery

47

Blood vascular systems

In most cases the contents of the piped system is blood, the composition of which varies.

Plasma is the watery medium in which the various solid and soluble materials are found. It is in and through plasma that the various body secretions, digestive products, waste and hormones pass, and thus it can be appreciated why mammalian physiology goes to such lengths to maintain a constant volume and constant chemical equilibrium within the plasma.

Corpuscles vary in both structure and function. The **erythrocytes**, or red blood cells, carry the respiratory pigment haemoglobin. The role of all respiratory pigments is the same, namely to form a loose association with oxygen in regions of high oxygen concentration, and to give it up when passing through regions of low oxygen concentration. It is found that the respiratory pigments in corpuscles carry more oxygen than pigments 'loose' in the plasma. The only problem is that corpuscular respiratory pigments require a certain minimum oxygen concentration before they can take up oxygen for carriage. Plasmic pigments may form an association at much lower concentrations.

The **leucocytes**, or white corpuscles have two main functions, carried out by specialized types of leucocyte. Firstly, they produce antibodies – chemicals that counteract poisonous compounds from bacteria. Secondly, there are some leucocytes that engulf bacteria, in the same fashion as *Amoeba* eats its food, thereby eliminating potential sources of infection.

Thrombocytes aid in clotting of blood thus stopping the loss of valuable water and materials, at the same time blocking entry of pathogens.

A piped supply of blood is known as a closed system; the unpiped form found in arthropods is known as an open system. Closed systems may be single or double cycles. A single cycle means the blood flows through the heart once in a complete circuit, namely: heart → respiratory organ → body → heart.

In the double cycle the blood passes through the heart twice in any complete circuit: heart → lungs → heart → body → heart. The double cycle is seen in mammals. The separate components of a double cycle permit different blood pressures for the lungs and for the rest of the body.

(*Top*) Blood circulation in a mammal. (*Bottom*) Blood circulation in a fish. Arrows indicate the direction of flow.

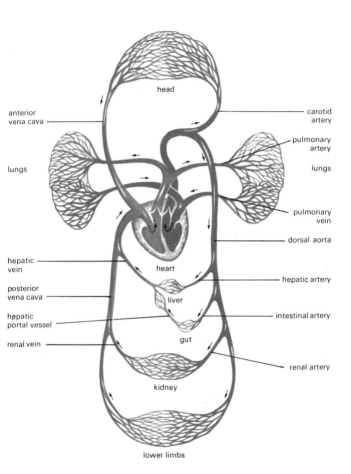

head

anterior vena cava

carotid artery

lungs

pulmonary artery

lungs

pulmonary vein

dorsal aorta

hepatic vein

posterior vena cava

hepatic portal vessel

renal vein

heart

liver

hepatic artery

intestinal artery

gut

renal artery

kidney

lower limbs

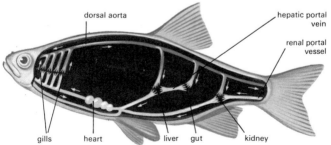

dorsal aorta

hepatic portal vein

renal portal vessel

gills   heart   liver   gut   kidney

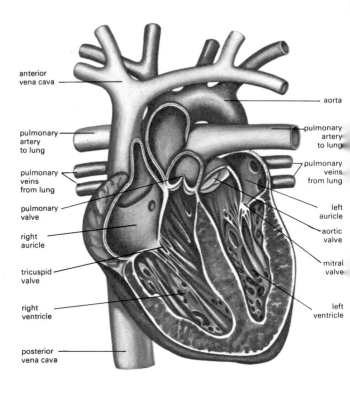

anterior vena cava

aorta

pulmonary artery to lung

pulmonary artery to lung

pulmonary veins from lung

pulmonary veins from lung

pulmonary valve

left auricle

right auricle

aortic valve

tricuspid valve

mitral valve

right ventricle

left ventricle

posterior vena cava

(*Above*) Mammalian heart partly cut away to show the structure.
(*Below*) The pumping cycle of the heart.

auricles fill with blood (diastole)

ventricles force blood out (systole)

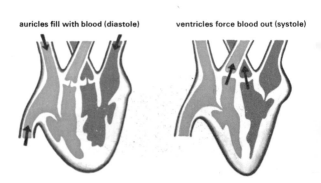

Pumping mechanisms

For an efficient blood system, there must obviously be some means of propulsion. One of the simplest methods is simply contractile vessels squeezing the blood along, as seen in many worms. Although the worm can be said to exhibit division of labour in having a piped blood supply, there is no one organ responsible for blood propulsion.

Many animals exert pressure on the blood flow when they move. This is true of a very wide range of animals from worm to man. A soldier standing motionless, at attention, can assist the return of blood to the heart by contracting his leg muscles, the pressure on the veins helping to push blood along. The flexure of the dogfish tail compresses segmental capillaries, but the main tail vessels are protected within a canal in the cartilaginous vertebrae – here the blood flow has been isolated from the major propulsive body movements.

Hearts

Hearts vary enormously in structure and size, from the simple valved tubes of worms to the multi-chambered organ of vertebrates. We shall consider only vertebrate hearts generally, and mammalian hearts in detail.

All vertebrates' hearts consist of several chambers, the chief function of which is to build up blood pressure. In any one complete circuit of the body, the blood passes through the heart once or twice depending on the class of vertebrates concerned.

Within the vertebrates there is a general trend towards a separation of functions within the heart, into a right side concerned with reception of deoxygenated blood from the body and its despatch to the respiratory surfaces, and the left side which receives and propels oxygenated blood round the body. The fish show no such division, there being a linear arrangement of four chambers concerned solely with receiving blood, at a very low pressure, and its despatch onto another circuit.

The amphibian heart, having two auricles and one ventricle, has been the subject of some controversy, many authorities believing it to be a degenerate four-chambered structure. Deoxygenated blood is received by the right auricle, and oxygenated blood from the lungs, by the left auricle. The single ventricle, however, appears to have some mixing of blood occurring within it.

In reptiles, mammals and birds, the heart is separated into a right pulmonary half, and a left systemic half.

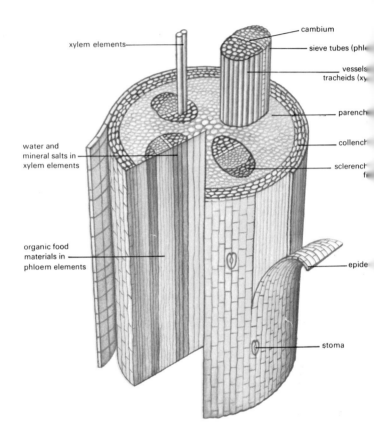

xylem elements

cambium

sieve tubes (phle[...]

vessels
tracheids (xy[...]

parench[...]

collench[...]

sclerench[...]
fi[...]

water and
mineral salts in
xylem elements

organic food
materials in
phloem elements

epide[...]

stoma

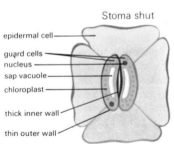

Stoma shut

epidermal cell

guard cells

nucleus

sap vacuole

chloroplast

thick inner wall

thin outer wall

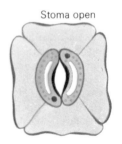

Stoma open

Transport in plants

All plants above the evolutionary level of the mosses and liverworts possess vascular tissue for the piped transport of water, salts and sugars. There is a division of labour, the xylem transporting the water and mineral salts upwards from the soil, whilst the phloem carries organic foods. Phloem and xylem are grouped into combined *vascular bundles*.

Animal vascular systems have a pump, the heart, but no such structure exists in plants. Instead the major force taking water along and up the xylem vessels is atmospheric pressure allied to the evaporation of water from the leaves. Mention was made on pages 8 and 9 of the small openings (stomata) which permit flow of gases for photosynthesis. At the same time there is a considerable loss of water via stomata. The enormous surface area of leaves on a tree permits the evaporation of hundreds of gallons per day. This water loss results in a pull on the water column in the xylem; as the top of the column in the leaf evaporates it is replaced by molecules of water from the xylem. The forces of cohesion between the water molecules in the xylem are sufficient to maintain a continuous column of water.

Evaporation of water from a leaf is called *transpiration* and it can be seen that this burden on the plants is put to good use in two ways: firstly, by maintaining a flow of water and salts; and secondly, by cooling the leaves due to evaporation of water.

The stomata normally open in the day and close at night – the mechanism for this process causing some controversy between different authorities. One explanation depends on the fact that the conversion of starch to sugar occurs faster in alkaline conditions, or at least, in regions of low acidity. The converse is true, namely sugar to starch conversion occurs faster in acidic regions. These reactions are catalysed by acid-sensitive enzymes.

At night carbon dioxide collects, due to respiration, and creates a concentration of carbonic acid. The acidity favours starch formation in the guard cells of stomata. At dawn, photosynthesis starts and the acidity drops as the carbon dioxide is used. This drop in acidity favours starch to sugar conversions. The rise in sugar concentration results in a rise in osmotic pressure, and water flows into the guard cells. The rise in turgor pressure within the guard cells causes the walls to stretch, the differential thickening results in a 'gaping' of the stoma.

(*Top*) Section of dicotyledonous stem.

COVERING LAYERS

The limiting layers of both plants and animals serve the same functions and must solve much the same problems. Basically, both must aid in retaining the general shape, and at the same time protect the organism.

Shape retention

This function can be achieved either by physical strength of inflexible armour plating, as seen in the exoskeletons of arthropods, e.g. crabs, or by elastic fibres, shown very well in mammals. Plants do not have quite the same problem as each cell of a plant has a rigid cellulose cell wall and hence the overall shape is governed by growth of the various parts, rather than by a constricting layer.

Protection

Physical protection against predators and the environment, is seen in both plants and animals. The arthropod exoskeleton serves as armour, the sheer bulk giving protection. Mammals have layers of dead and dying cells, the outermost layer of dead flattened cells giving a tough barrier to most abrasions. In addition, most mammals have a dense fur which not only protects, but also camouflages.

Many plants have a layer of bark which can withstand blows, and also prevents entry of pathogens to the innermost regions.

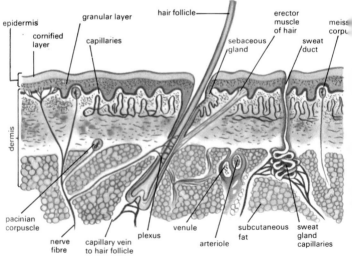

Even plants lacking bark frequently have a wax cuticle covering flattened epidermal cells.

Terrestrial organisms have the problems of water balance generally, and water retention especially. In most cases terrestrial animals and plants have developed an impermeable layer, e.g. wax cuticles in plants, dead cell layers in mammals and mixed protein and fat layers in insect cuticles.

It should be pointed out, however, that the water retention properties of both mammalian and insect coverings alter with increase in temperature. As mammals become heated by sun and activity, sweat is produced to cool them by evaporation from the surface. As the temperature rises many insect cuticles start to lose water. This is due to rearrangement of the wax molecules in the waterproofing at a certain temperature – the critical temperature.

(*Opposite*) Vertical section of mammalian skin. (*Below*) Vertical section of plant cuticle. (*Bottom left*) Detail of stoma. (*Bottom right*) Vertical section of arthropod exoskeleton.

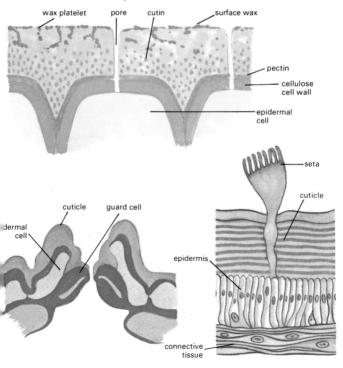

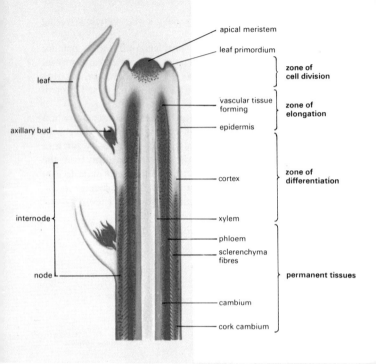

apical meristem

leaf primordium

zone of cell division

leaf

vascular tissue forming

zone of elongation

axillary bud

epidermis

zone of differentiation

cortex

internode

xylem

phloem

sclerenchyma fibres

permanent tissues

node

cambium

cork cambium

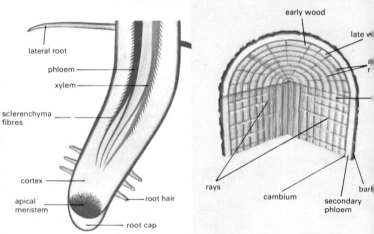

lateral root

early wood

late wood

phloem

xylem

sclerenchyma fibres

cortex

rays

root hair

cambium

apical meristem

root cap

secondary phloem

bark

GROWTH

To say an organism grows is a statement immediately understood by most people, as they think they are well aware of what is meant. If, however, one attempts a definition of growth, there are problems.

Some might say growth is an increase in weight, but is this wet weight (total weight), or the weight left after water has been removed? Many germinating seeds increase in weight, but this is due solely to uptake of water, and if such seedlings are selected and dried, it will be found that their dry weight has dropped as dry stored food material has been utilized in germination. Similarly newborn children may lose weight for the first few days, but are they not growing?

Other authorities have said growth is accompanied by an increase in cell numbers, but many stages of plant growth, where there has been an obvious increase in length of some structure, can be shown to be due to elongation of existing cells and not an increase in numbers.

This last case points to another way of determining growth, namely change (increase) in overall size of the organism. This can usually be taken as growth, but it should be remembered that many dry seeds will swell (grow?) when soaked in water. Similarly the unfolding wings of newly hatched butterflies increase in size as air and blood are pumped into them.

The points above indicate the problems of terminology and account for the numerous techniques devised to measure the different aspects of growth.

Meristems of plants

Plants typically grow in length throughout their lives, and the regions concerned with the division and increase in numbers of cells are the meristems. The meristem may be a single cell at the ends of the plant body, e.g. liverworts, or it may be a multicellular patch of cells all actively dividing, as is common in many higher plants.

The diagrams show the basic plant form, and a comparison of basic structure with the plan of meristems indicates the sites of cell division. The cambium of a vascular bundle is merely an extension from the unspecialized dividing cell tissue of the apical meristem. Many cambia are produced later in the life of the plant in response to a change in conditions, for example, wound bark cambia.

(*Top*) Vertical section of stem. (*Bottom left*) Vertical section of root. (*Bottom right*) Section of mature tree trunk.

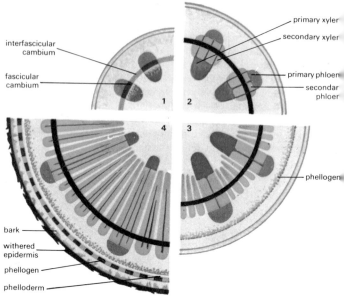

interfascicular cambium

fascicular cambium

primary xylem

secondary xylem

primary phloem

secondary phloem

phellogen

bark

withered epidermis

phellogen

phelloderm

Sequence of tissue development during secondary growth.

Secondary growth in plants

As plants grow, competition for space and problems of reproductive display become acute. Many, especially flowering plants and conifers, solve the problem by growing taller, as this enables leaves to gain light whilst displaying reproductive organs to wind and insects. Increase in height brings transport and support problems, however, solved by the addition of material to the stem, thereby increasing the physical bulk and strength of the plant. This addition is known as *secondary thickening*.

The four-quadrant diagram shows the basic sequence of events. Fascicular cambium in the vascular bundle becomes active, cutting off secondary phloem and xylem. Fascicular cambium is joined in its activities by the intervening cells of the cortex becoming meristematic and forming interfascicular cambium. Once these two cambia link, it is not long before a continuous cylinder of secondary phloem and xylem forms. The secondary xylem is always more obvious and forms the bulk of the secondary tissue. It is secondary xylem that forms commercial wood.

The increase in girth of the stem stretches the outermost tissues, with the possible risk of splitting of the epidermis and outer cortex. Such splits would permit the entry of bacteria and fungi. To prevent this, a layer of cells in the outer cortex becomes meristematic, cutting off regular rows of box-like cells impregnated with a fatty material. These cells are corky bark tissue produced by a cork cambium.

At intervals the barrier of bark is disturbed due to an irregular arrangement of cells permitting flow of gases to outer tissues. Such disturbances are lenticels or breathing pores (see illustration).

To permit diffusion of gases radially through the secondary tissue there are rows of unspecialized cells (medullary rays) cut off again by the cambia.

The production of secondary tissue occurs mainly between spring and autumn. During spring the newly formed secondary xylem has large thin-walled cells; as autumn approaches the vessels formed are smaller with thicker walls. This seasonal influence is seen in a transverse section of a stem as a series of rings (*annual rings*). Thus the age of a tree may be estimated. A similar secondary development is seen in roots as they too require greater strength as the aerial regions develop.

The above account of secondary thickening is true of dicotyledonous plants, but not so of the monocotyledonous plants such as cereals, grasses, rushes and reeds. Such plants usually rely on thickening of the cellulose walls of the outer cortex to provide a tough yet flexible cylinder of supporting tissue.

Lenticel

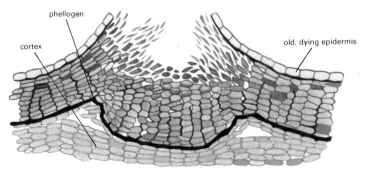

phellogen

cortex

old, dying epidermis

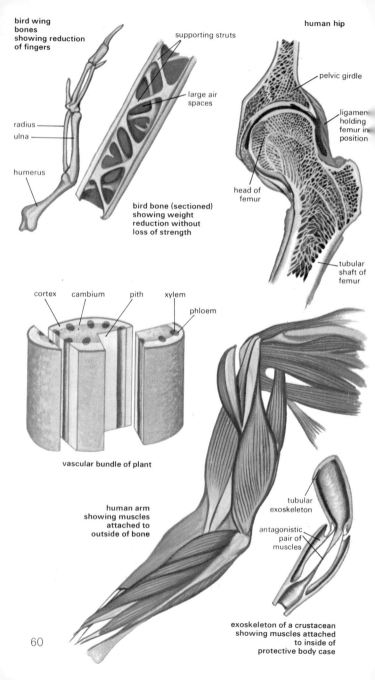

bird wing bones showing reduction of fingers

radius
ulna
humerus

supporting struts
large air spaces

bird bone (sectioned) showing weight reduction without loss of strength

human hip

pelvic girdle
ligament holding femur in position
head of femur
tubular shaft of femur

cortex
cambium
pith
xylem
phloem

vascular bundle of plant

human arm showing muscles attached to outside of bone

tubular exoskeleton
antagonistic pair of muscles

exoskeleton of a crustacean showing muscles attached to inside of protective body case

Strength in plants and animals

The trunk of a tree has two major forces acting upon it: the weight of the tree which acts vertically, and the more damaging horizontal forces due to wind pressure. In some recent experiments at the Royal Aircraft Establishment, Farnborough, England, it was found that the lateral forces of wind pressure were equal to the weight of the tree at 40 miles per hour. Obviously trees are exposed to gusts of wind greater than 40 miles per hour and therefore they have to cope with lateral forces greater than their own weight.

When a lateral force pushes on a tree the exposed side is stretched, whilst the sheltered side is compressed. It has been found that the best shape for an object subjected to such forces is tapering, with the thickest part at the base. Furthermore, the ideal cross-section is circular. These last two points account for the shape of most of the trees alive today, and is exemplified superbly by the tallest specimens, the giant redwoods.

The conflicting factors found in growth are, a limited supply of building materials tending to encourage as slender a stem as possible, whilst the physical problem of lateral stresses demands as thick a trunk as can be produced.

Young stems and roots show a great difference in distribution of vascular tissue. The stem has basically a cylinder of vascular bundles, the root a solid central mass of tissue. The explanation lies in the type of forces acting on them. In a root the forces are, tugging from above due to swaying of the aerial portions, allied to the need for flexibility in order to grow around underground obstructions.

The stem, as already mentioned, needs great flexible strength to withstand the drag forces of wind pressure.

The long bones of animals have similar problems to stems – great lateral forces acting on them when a limb moves, plus the possible problem of bearing the weight of the animal from above. There is no such root problem of tugging with enormous flexibility. Since the problems are similar to stems it is no surprise to find the same solution has evolved – cylindrical structures tending to have hollow centres, the hollow cylinder giving much greater mechanical flexible strength than a solid rod. This last feature can be easily compared and illustrated in the use of hollow, tubular scaffolding where lateral stresses have to be endured.

Cylindrically shaped structures are widely used throughout both the animal and plant kingdoms to provide strength and support.

A

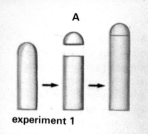

experiment 1

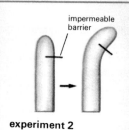

impermeable
barrier

experiment 2

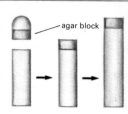

agar block

experiment 3

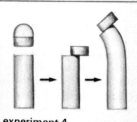

experiment 4

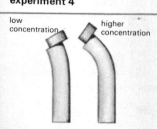

low
concentration

higher
concentration

experiment 5

B

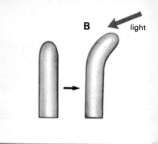

light

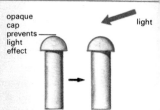

opaque
cap
prevents
light
effect

light

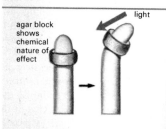

agar block
shows
chemical
nature of
effect

light

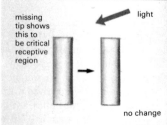

missing
tip shows
this to
be critical
receptive
region

light

no change

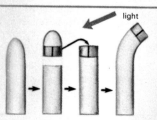

light

Plant growth control

Obviously there must be some system governing the growth of plants as they have a characteristic form typical of their species. Controlling factors on the rate of growth would appear to be such features as sunlight for photosynthesis, availability of raw materials, etc. The shaping influences of plant growth are also external stimuli, such as light and gravity.

If a plant is placed on a window sill there is the well-known response of bending towards the light, exposing the leaves to advantage. Such a growth response of part of a plant, in response to an external stimulus, is called a *tropism* (in the case of the example quoted, positive phototropism, i.e. bending towards a light stimulus). Other tropic responses are geotropism (gravity) and hydrotropism (water). Detection of the various stimuli occurs in different regions, light being detected by the growing shoot tip – shown by covering the tip with an opaque cap when no positive phototropism is forthcoming. Similarly the tip of the root is the gravity-sensitive region.

In illustration A, experiment 1 shows that a shoot tip exerts some influence over the growth of a shoot; its removal stops growth, its replacement causing growth to continue. Furthermore, experiment 3 indicates that the tip's influence is chemical rather than physical, as removing a tip, placing it on a block of agar jelly for two hours and then placing the jelly on the stump initiates further growth. Obviously the only connection between tip, agar block and stump must be in some chemical passing from tip to agar block and then from block to stump.

Experiment 2 shows what happens when some physical barrier is placed in the way of downward flow of chemicals: This indicates that the side having the chemical along its length grows faster. Experiment 5 shows that the amount of chemical influences the amount of curvature. The chemical has been isolated and identified as a plant hormone, or auxin, called indol acetic acid.

Illustration B describes a range of experiments on phototropism, again showing auxin effects.

Movement of an entire organism, in response to external stimuli, is called a *taxis*, and is seen in unicellular algae which exhibit positive phototaxis, i.e. they move towards light, aiding their photosynthesis. Diffuse stimuli, such as temperature and humidity, result in nastic movements, e.g. opening and closing of flowers.

Experiments showing the effects of hormones on plant growth.

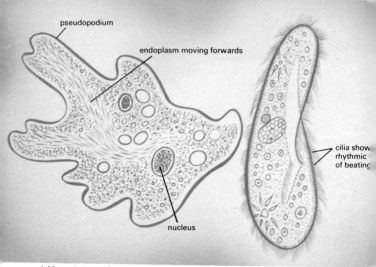

(*Above*) Locomotion in *Amoeba* (left) and *Paramecium* (right). Both the jellyfish and *Nautilus* (below) move by jet propulsion, forcing water out of the body cavity.

MOVEMENT

Movement is a characteristic of most animals (and some plants), but having said that, one is left with a range of methods of loco-motion, each method often having exponents drawn from widely different groups of animals.

Cytoplasmic flow This is seen in the formation of pseudopodia, well known in *Amoeba*. Here the innermost, more liquid endoplasm seems to flow forwards causing the more viscous ectoplasm to bulge forwards. This flow of endoplasm is possibly due to a change in state of the gelatinous ectoplasm at the posterior, the ectoplasm constricting and pushing the endoplasm forward. Pseudopodia can be formed anywhere on the body surface, hence movement does not involve a permanent specialized region.

Cilia and flagella These are extensions of a cell and are common methods of locomotion at the protistan level of organization. Their regular beating assists the organism not only to be rowed along, but frequently induce a feeding current. Cilia and flagella are permanent specialized organelles.

Body movements Many animals do not have specialized locomotory organs, such as legs or wings, but rely on wriggling of the body or changes in body shape. Movement is often aided by extensions of the body surface, in order to push against either water or land. This feature is very well illustrated by the paddle-like extensions (parapodia) of some marine bristleworms, e.g. the ragworm. In most of these types, the passage of rhythmic waves of muscle contraction along the body causes a resultant equal and opposite reaction by pushing against the substrate. This is shown in such land animals as snails, creeping along on the muscular foot and snakes sinuously pushing along.

Liquid skeleton All muscles need something against which to pull when they contract, the bone skeleton of vertebrates being an example. Many animals, however, lack a solid skeleton, whether internal or external, and such creatures frequently rely upon a volume of incompressible body fluid as a mass upon which to pull, frequently the fluid of the body cavity, e.g. the earthworm.

Jet propulsion Some animals take in a volume of liquid and then force it out of some cavity via a small opening, producing a jet of liquid which forces them forward, e.g. squid, jellyfish and *Nautilus*.

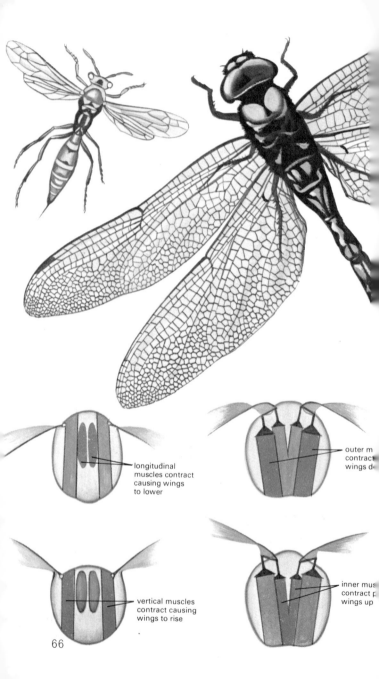

longitudinal
muscles contract
causing wings
to lower

outer m
contrac
wings d

vertical muscles
contract causing
wings to rise

inner mus
contract p
wings up

66

Exoskeletons

The arthropod exoskeleton performs not only the usual skeletal functions of support, system of levers, protective case, source of calcium, etc., but has proved to be capable of great variation in form, despite the well known problems of weight, restriction in growth, and the need for several joints per limb.

Typical walking legs of arthropods are seen in the cockroach, leaping legs in grasshoppers, whilst some limbs play dual roles, e.g. walking limbs of crustacea may well have gills at their base.

The insects typically have two types of appendage: three pairs of walking legs and two pairs of wings – one pair for each of the second and third thoracic segments. The wings of insects have a long evolutionary history, and are thought to have evolved as fixed lateral gliding devices that were there to increase the surface area in the heaviest region of the body, the thorax. Only two pairs exist in living forms, and in the true flies the second pair are modified as balancing organs. In beetles, and to some extent the cockroach, the first pair of wings form a wing case or covering for the membranous second pair. The wings may be operated by two sets of muscles: **direct muscles**, attached to the base of the wing and therefore pulling directly on them, responsible for altering the angle of wing; and **indirect muscles**, comprising the dorso-ventral muscles attached to the roof (tergum) and the floor (sternum) of the thoracic skeleton, plus the longitudinal muscles running from front to back of the segments and attached to compartments dividing one segment from another.

The indirect muscles cause changes in the shape of the thoracic segments, resulting in wing movements up and down, the contraction of one set of muscles stretching the antagonistic pair which in turn contract, thereby stretching the others, and so on.

Some insects, e.g. butterflies, have relatively slow wing beats – 10–40 beats/second, and have a nerve impulse for each wing beat. In those with much faster wing beats, e.g. houseflies – 200 beats/second, mosquitoes – 600 beats/second and some midges – 1000 beats/second, the rate is faster than direct nerve stimulation can achieve. This point puzzled biologists until they discovered the antagonistic 'triggering' method of indirect muscle action.

Cross sections through the thoracic regions of (*left*) a wasp to show indirect flight muscles and (*right*) a dragonfly to show direct flight muscles. In each case the muscle which is in a state of contraction is drawn red.

The evolution of limb position and posture during the conquest of land by the vertebrates.

Land colonization

The transition from life in water to life on land posed several problems, as listed below.

Support Life on land in an air medium puts greater strains on the body than living in water does.

Locomotion On land the typical aquatic locomotory organs of cilia, paddle-like limbs, or propulsive tail, are useless. Animals need either limbs to lift them clear of the ground, or, if they leave the body fully in contact with soil, must rely on specialized body movements, as are found in worms, snakes and snails for instance.

Respiration Use of the body surface necessitates a permeable moist skin; gills are useless as they desiccate, and the filaments stick together in the absence of supporting water. The adpressed gill filaments have a reduced surface area/volume ratio and are therefore useless. Respiration on land needs some internal tubular, sac-like organ, like a lung, or tracheole.

Dehydration This is a major problem as obviously aquatic forms do not have this worry; an impermeable skin is required.

Transfer of gametes No longer will external fertilization be possible and, to prevent death of gametes due to exposure, some form of copulation may well be required.

Change of sense organs In water, pressure-sensitive devices are required, whilst on land, sight is more important, although a modified ear is important also.

Excretion On land there will be a change from the possible production of ammonia, which is very soluble and hence easily coped with in water, to a less dangerous excreta, on land, e.g. urea and uric acid. Also, the production of urea and uric acid are water conservation devices, compared to ammonia which needs large volumes of water for its safe disposal.

Vertebrates and land locomotion

The diagrams of various vertebrate stages in the colonization of land are simplified, but show the probable sequence from a fish-like form to a land-going form. The early stages show how fins would be impractical on land both in size, orientation and also musculature. This development of a jointed limb from a fin is derived from fossils, but it should be emphasized that the changes in muscle size and orientation are equally important if the limb is to be of value.

The basic limb of all tetrapod (four-limbed) forms is the pentadactyl limb, ending in five digits. This has become modified greatly, to perform a variety of functions.

Adaptive radiation of the pentadactyl limb.

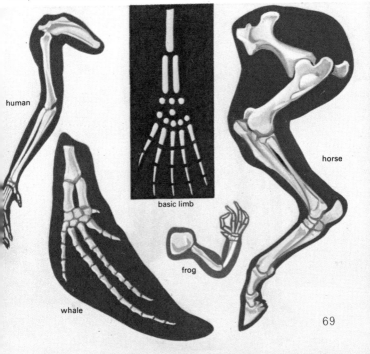

human

basic limb

horse

frog

whale

69

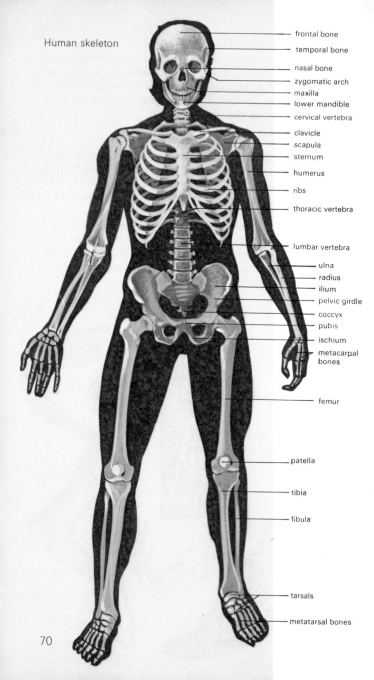

Human skeleton

frontal bone
temporal bone
nasal bone
zygomatic arch
maxilla
lower mandible
cervical vertebra
clavicle
scapula
sternum
humerus
ribs
thoracic vertebra
lumbar vertebra
ulna
radius
ilium
pelvic girdle
coccyx
pubis
ischium
metacarpal bones
femur
patella
tibia
fibula
tarsals
metatarsal bones

The human skeleton

An illustration of the human skeleton is a familiar picture to many people, yet its functions are often not known, but merely taken for granted. The functions of the skeleton are listed below.

Support This is the most obvious role, and the shape of the human form is based on its bony skeleton. Posture of the body is achieved by muscles pulling on the various bones.

Locomotion All tetrapods rely on the limb bones to provide a system of levers upon which muscles can pull to move the animal along. The joints enable bone to bone mobility without any excessive friction.

Site for muscle attachment This has been mentioned in passing, and the various outgrowths found on bones, particularly the vertebrae, are indications of the many muscle attachments. This combination of muscle and bone enables us to eat, move and breathe, etc.

Protection Many delicate and vital organs are sheltered from the outside world by parts of the skeleton: the cranium protects the brain, ribs protect the heart and lungs and vertebrae enclose the spinal cord.

Red blood cell formation The red blood cells (erythrocytes) are formed in the marrow of the long bones such as the femur. Any disease affecting the marrow will be of great importance in the erythrocyte economy of the body.

Respiration The ribs play a vital part in the breathing movements which result in air moving in and out of the lungs.

Calcium store Calcium is required for normal body metabolism, e.g. in blood clotting. In pregnant women, there is the added burden of taking in sufficient calcium ions for the developing young within the womb. Should the woman's diet be deficient in this element, calcium is eroded from her skeleton and passed to the foetus.

Joints

These sites of bone to bone articulation allow for movement in a variety of ways. A ball and socket joint such as is seen between the hip and the head of the femur, or the shoulder and the head of the humerus, provides for movement in three planes. The hinge joints of the knee and elbow permit a controlled movement in one plane only. A swivel joint between the top two vertebrae – the atlas and axis – allows a controlled rotational movement seen in shaking of the head. Sliding joints (modified forms of ball and socket) are found in wrists and ankles. Finally, there are the fixed joints which no longer permit movement, e.g. skull bones.

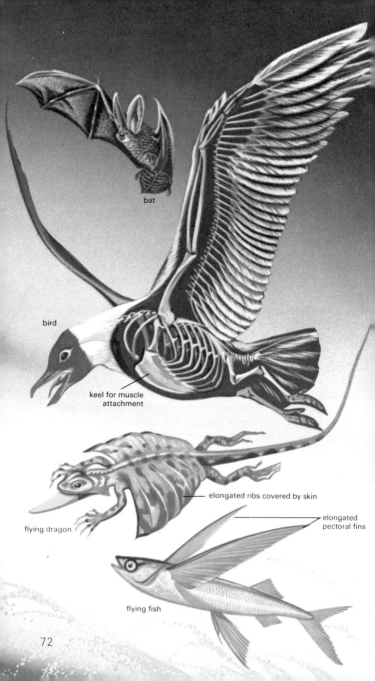

bat

bird

keel for muscle
attachment

flying dragon

elongated ribs covered by skin

elongated
pectoral fins

flying fish

72

Flight

The ability to fly has evolved in two phyla of animals, the arthropoda and the chordata. In addition there is one class of molluscs that show some ability to glide. Flight can be considered from the means of propulsion used.

Parachuting A parachute is prevented from falling straight to the ground under the influence of gravity by developing an aerodynamic force (drag force) which balances its weight. The drag force is generated by the air stream moving parallel upwards with respect to the parachute's vertical descent. Drag force depends on three factors: the surface area of the parachute, the density of air and the speed of relative air flow. To achieve the best parachute effect an organism should be radially symmetrical – an unusual feature in freely moving organisms. Such a perfect state is achieved only by some fruits and seeds, e.g. the dandelion, and this accounts for the effectiveness of dispersal. The only drawback to wind dispersal is its passive nature.

Gliding This movement is achieved by bilaterally symmetrical forms which deflect the air flowing past them and, in a straight glide, the weight is balanced by the net aerodynamic force. Gliding can only be maintained by a net loss of height if speed is to remain constant or be increased. Occasionally parachuting may be the result of a specialized short term glide.

Gliding is known in every class of chordates, e.g. flying fish have large pectoral fins which, after breaking through the surface of the water, enable them to escape predators by gliding. Many mammals, both marsupial (phalangers) and placental (North American squirrels), use a fold of skin stretched between fore and hind limbs. The flying lizard, *Draco*, of the East Indies has a fold of skin supported by elongated ribs. The flying frogs of Borneo use enormously extended webs between elongated toes.

Wings A group of fossil reptiles, the pterosaurs, had a wing formed by a fold of skin extended between fore and hind legs, the bulk of the front support being an extended fourth digit. This wing was inefficient and any tear would render it useless. Bats' wings are formed from a flap of skin between elongated fingers, plus an extension back to their legs and tail. Birds use their entire forelimb as a flight surface and the tail for steering.

Flight in four different classes of vertebrates – an example of convergent evolution. Only the bat and the bird exhibit flapping flight, however.

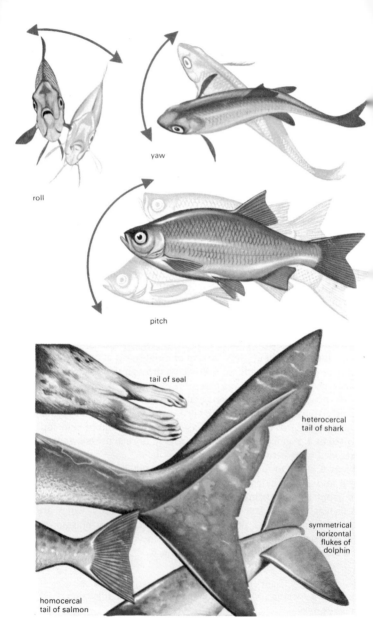

roll

yaw

pitch

tail of seal

heterocercal tail of shark

symmetrical horizontal flukes of dolphin

homocercal tail of salmon

Movement in water

Water has a high relative density, which means that it is a relatively viscous medium offering great resistance to any object attempting to move through it. To overcome this, streamlining is necessary. As compensation for the viscosity, water does provide support for bodies in it, and its viscosity means an animal can use appendages to push against it, in order to gain propulsion.

The best known aquatic animals are fish which show a wide range of form and function. Fish may be divided into two groups, based on certain anatomical features: the bony skeleton fish, and the fish with a skeleton made of cartilage. The bony fish have another important structure, namely the air sac, or *swim bladder*, situated on the dorsal wall of the abdominal cavity.

The swim bladder is an outgrowth of the pharynx and enables a fish to change depth by altering its density. In some fish, e.g. goldfish, the swim bladder retains its connection with the pharynx, and thus the fish can sink by 'blowing bubbles' – a feature which increases its density. The problem with this open type of swim bladder is that to refill with air, a trip to the surface is necessary. The closed swim bladder relies on gas being secreted into it, or withdrawn from it, using glands in the walls of the swim bladder itself. The dorsal position of the swim bladder aids in keeping the fish on an even keel. The cartilaginous fish lack a swim bladder and tend to sink, as they are denser than water. This downward movement is only averted by constant forward movement. Forward movement can be achieved by active propulsion from the tail, or by merely using the downward force of the fishes' own greater density. The shape of the fish plus the use of its pectoral fins enable it to glide through the water.

The tail shape can affect up and downward movement (pitch) in a fish by pushing the tail region up, as, for example, in the asymmetrical tail of the dogfish. The diagram shows the two main shapes of fish tails: symmetrical about the longitudinal axis in the vertical plane (homocercal), found in the bony fish; and the asymmetrical heterocercal tail typical of the cartilaginous fish. The tails of aquatic mammals, such as the whale, porpoise and dolphin are symmetrical but lie in the horizontal plane.

(*Top*) The main propulsive force is produced by the tail of a fish. Stability is maintained by the median fins (dorsal, anal and ventral fins) which correct roll and yaw, and by the paired fins (pectoral and pelvic fins) which correct pitch. (*Bottom*) Tails of different aquatic vertebrates.

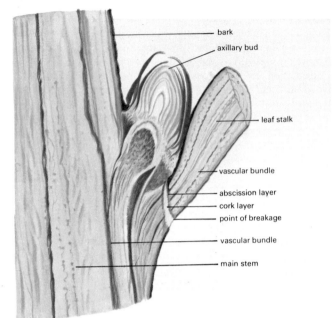

bark

axillary bud

leaf stalk

vascular bundle

abscission layer

cork layer

point of breakage

vascular bundle

main stem

(*Above*) Leaf fall in sycamore. (*Below*) Exudation of excess water from a leaf occurs in conditions of high humidity.

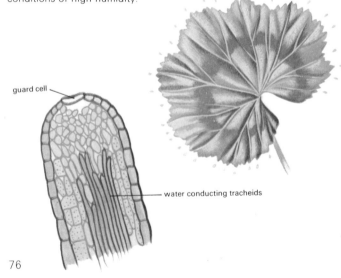

guard cell

water conducting tracheids

EXCRETION

The many chemical and physical activities of an organism result in numerous waste materials, and it is the elimination of these materials that have once been part of the organism, i.e. have actually passed through a cell membrane, that is called excretion. Excretory products are of two main types.

Carbonaceous This is waste involving some carbon compound, e.g. carbon dioxide. The carbon dioxide is usually excreted via the respiratory surface (see page 45), just occasionally it is recycled as in the building of calcium carbonate in the crustacean shell or in the calciferous glands of earthworms which aid in neutralizing acidic food.

Nitrogenous The activities of plants and animals result in the 'wearing out' and degradation of a variety of cell parts, many built from protein. The simplest nitrogenous excretory product is ammonia, and, as has been mentioned earlier (page 68), this is disposable without further change, if the organism is relatively small and provided with a constant flow of water. The great solubility of ammonia proves of great value in this situation. Ammonia is extremely toxic, however, and animals evolving on to land were successful wherever they could convert the lethal ammonia to some less toxic nitrogenous waste. Two solutions to the problem were the formation of either urea or uric acid. In urea formation two molecules of ammonia combine with one molecule of carbon dioxide to form the relatively harmless urea molecule. This has the advantage of using up waste carbon dioxide. The disadvantage of urea excretion is the necessity for water to form urine, although not requiring as much water as ammonia. The formation of urea occurs in the liver, its elimination in the kidney being described on page 79.

Uric acid is very useful as it may be excreted as a solid paste thereby saving water, and hence is found in weight and water-conscious animals, e.g. birds and desert animals. It is also excreted by land snails and insects.

Plants also produce waste carbon dioxide from respiration, but it only becomes apparent during darkness, as photosynthesis utilizes all of it during the day. Leaf fall enables many plants to rid themselves of waste material, nitrogenous waste included, the event being annual in deciduous forms, and all year round in conifers. Leaf fall also removes the great surface area of leaves that would provide wind resistance in winter. Some nitrogenous waste is built into the secondary tissue developed during the growth of plants.

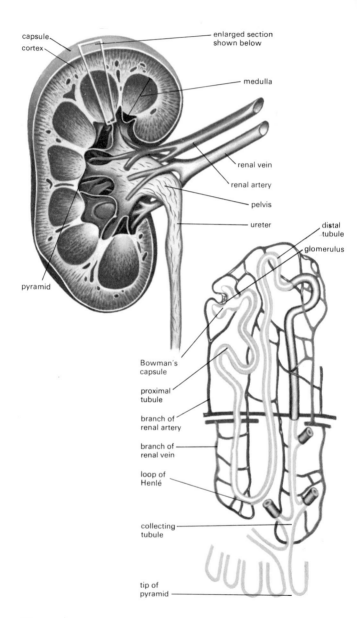

capsule

cortex

enlarged section
shown below

medulla

renal vein

renal artery

pelvis

ureter

distal
tubule

glomerulus

pyramid

Bowman's
capsule

proximal
tubule

branch of
renal artery

branch of
renal vein

loop of
Henlé

collecting
tubule

tip of
pyramid

Excretion in vertebrates

Blood arriving at the vertebrate kidney contains numerous materials, both useful and waste, and furthermore the blood pressure is high. It is this hydrostatic pressure that forces materials out through capillary walls in the glomerulus, and into the cavity of the walls of Bowman's capsule. The only structures retained in the capillary are those with a high molecular weight, e.g. erythrocytes, some blood proteins, plus a little plasma.

Materials forced into the capsule now pass along the length of the nephron, being kept moving by the pressure of more materials arriving. As materials pass through the coils of the *proximal convoluted tubule* there is considerable reclamation of water, glucose and amino acids. At one time it was thought that the viscous blood, flowing in the capillaries around the tubules, exerted some osmotic force on the water in the convoluted tubules. It is now known that reclamation of materials from the proximal convoluted tubule into the capillaries is active. Three pieces of evidence suggest an active process. Firstly, lowering the oxygen concentration in this region of the tubules slows the rate of transfer, i.e. active processes are related to oxygen concentration. Secondly, narcotics injected into this region slow down the rates of transfer, a sign of active reclamation. Thirdly, injection of materials into the convoluted tubules to make them isotonic (same osmotic pressure) with blood capillaries does not stop the flow, i.e. osmosis cannot be involved.

The efficiency of water reclamation can be judged by a consideration of how much water is filtered via Bowman's capsules, and how much is excreted. Out of 170 litres passing through Bowman's capsules, approximately 150–155 litres are reabsorbed in the region of proximal convoluted tubules.

The *loop of Henlé* provides a system for increasing the osmotic pressure of the medullary region, by the recycling of sodium. The movement of sodium ions is from ascending to descending limbs of the loop, and thus the intervening tissues have a high osmotic pressure. Sodium ion movement is under the control of the hormone aldosterone, produced by the adrenal gland.

What is the value of this concentration of sodium salts in the bottom of the loop of Henlé? The answer lies in an examination of the volume and quality of urine found at the top and bottom of the collecting duct, and will be considered overleaf.

(*Top*) Vertical section through a kidney. (*Bottom*) Detail of a nephric tubule.

The kidneys of rats from different environments showing the variation in the size of the medullary regions.

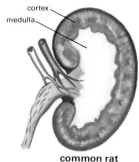

common rat

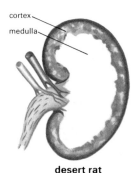

desert rat

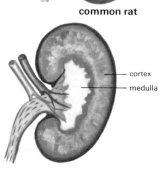

water rat

Urine arriving at the top of the *collecting duct* is relatively dilute, and hypotonic to blood (at a lower osmotic pressure). The urine passing into the pelvis of the kidney from the collecting duct is hypertonic to blood (at a higher osmotic pressure). The increase in concentration is achieved by selective withdrawal, or reclamation, of water. Water movement from collecting ducts to blood capillaries in the medullary region is due to that region having a very high osmotic pressure – the result of the aldosterone action in the loop of Henlé. Passage of water from the collecting duct is regulated by the variable permeability of the collecting duct walls, this being con-trolled by a pituitary hormone, vasopressin ADH (antidiuretic hormone). A diuretic is a chemical inducing a high excretion of urine, an antidiuretic therefore enables the body to conserve water.

ADH flow depends upon the amount of water in the body. In summer, when sweating and high evaporation rate result in great loss, the ADH concentration will be high in order to conserve water and thereby maintain an equilibrium within the body.

Factors influencing urine concentration

Length of loop of Henlé The greater the length, the greater the chance of any one sodium ion being recycled.

Rate of urine flow The faster the flow, the less time available for transfer of ions, i.e. the establishment of an osmotic gradient.

Rate of blood flow This affects the rate of movement of water and sodium ions across the medullary region, and therefore we find only one percent of renal blood in medullary capillaries despite being large capillaries, i.e. very slow blood flow.

Uric acid is an excellent excreta for any terrestrial organism. It is nontoxic and very nearly insoluble in water – a water conservation feature. It is excreted by many reptiles and birds, and seems to have evolved in vertebrates with the shelled egg – an asset where the developing embryo was going to grow up alongside its own waste, and where water was going to be at a premium.

Insect malpighian tubule – a factor contributing to their success.

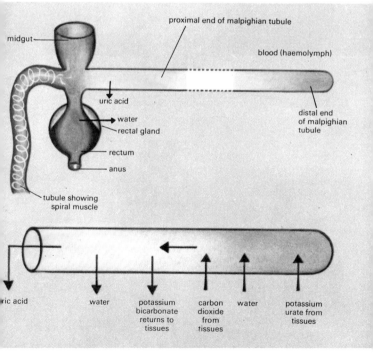

potato plant with tubers

runner of strawberry

bulb of daffodil

corm of crocus

rhizome of iris

REPRODUCTION AND DEVELOPMENT

As all organisms are destined to die, it is essential, for the continuation of the species, that they possess the powers of reproduction at some time during their life history. The methods of reproduction within the animal and plant kingdoms vary enormously, from simple division of the body into two halves, e.g. *Amoeba*, to the complex social and physiological events involved in the reproduction of mammals. If the production of the young is simply by the division of only part of one adult body then this is asexual reproduction, and can be seen very clearly in such organisms as yeast and *Hydra*.

The asexual production of young is a means of building up numbers of a species at a time that is favourable to the adult. This may have two drawbacks. Firstly, there is the immediate problem of possible overcrowding and competition followed by the likely death of some of the young – a very wasteful process. Secondly, there is the far more important long-term problem that the young will be identical, in their nuclear and chemical constitution, to the original parent. This could well result in the production of a large number of individuals all of which could be susceptible to the same disease. Hence, on the arrival of such a disease, there is a strong possibility of the whole community being killed. In a normal population, which is the result of many sexual matings, there is considerable variation in susceptibility to any disease and hence there could well be survivors of any infection.

Regeneration

Many organisms have the power of growing a replacement following the loss of an organ, or part of an organ. Primitive organisms are frequently capable of regenerating an entire organism from a very small piece of the original, e.g. mosses and sponges can be regenerated from a very few cells.

Perennating organs

An organ that continues the life of a plant from year to year is a perennating organ. Such structures frequently have the additional task of asexual reproduction. They enable the grower to guarantee the type of flower or crop which will result from the planting of the given perennating organ, whilst giving the chance of increasing the number of plants in the years to come.

Some perennating structures and asexual reproductive structures found in plants.

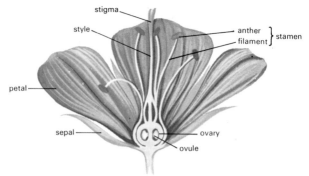

Vertical section through a flower to show the basic parts.

Sexual reproduction in plants

The essential feature of sexual reproduction in any organism is the act of fertilization – the actual fusion of the nuclei of the sex cells or gametes. One gamete, the egg or ovum, comes from the female, the other gamete, the sperm, from the male.

The production of gametes occurs within specialized organs, the gonads. The male gonad is the testis, the female gonad the ovary, and these gonads may be of only one sort per organism, i.e. the sexes are separate, or an organism may be *hermaphrodite*, i.e. it produces gonads of both sorts.

In animals there has been a trend towards a separation of the sexes, that is an animal is either male or female, the occasional hermaphrodite species, e.g. earthworm and *Hydra*, usually having some method of ensuring cross-fertilization.

Plants do not show the same picture of sexual separation as animals, the most successful group of plants alive – the angiosperms or flowering plants – being mostly hermaphrodite, or *monoecious* as it is called in plants. Nevertheless, although they produce both types of gametes, most flowering plants have methods, often quite complex, for ensuring cross-fertilization.

On page 68, mention was made of the problem of transfer of gametes, sperm in particular, in order that fertilization might be achieved. Some of the most primitive land plants alive today – the mosses and liverworts – still require external liquid water in which the male gametes actively swim to the ovum. During plant evolution, and the evolution of the land habit, there has been a suppression of this primitive external swimming of sperm, and the introduction of a tube along which the male gametes may pass.

The angiosperm flower

The flower is a structure, derived from modified leaves, having a number of different parts, each part having a basic function. The **calyx of sepals** protects the flower bud and may occasionally be coloured to attract insects, e.g. the anemone. The **corolla of petals** is the part of a flower used for 'advertising' purposes, attracting animals by its colour and possibly the scent of the sugary nectar. The **stamens** are the male parts of a flower producing pollen grains which contain the future male gamete and the **carpels** are composite structures containing ovules, each of which contains an ovum.

Some of the variations in reproductive structures found in green plants.

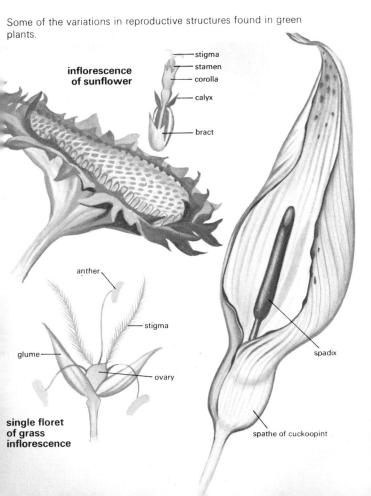

inflorescence of sunflower

stigma
stamen
corolla
calyx
bract

anther

stigma

glume

ovary

single floret of grass inflorescence

spadix

spathe of cuckoopint

Pollination

Pollination is defined as the transfer of pollen grains from the anther of a flower to the receptive surface of the stigma of the same, or another flower. When the pollen grain movements are within a single flower, it is termed self-pollination. Cross-pollination is the movement of pollen from one flower to another. Some authorities would qualify this by saying that cross-pollination is the transfer of pollen to the flower of another plant not merely from one flower on a bush to another flower on the same bush.

Clearly self-pollination with the possibility of self-fertilization is not as favourable, in evolutionary terms, as cross-pollination and cross-fertilization. Self-fertilization produces a zygote lacking the mixture of characteristics found in the products of cross-fertilization.

The transfer of pollen grains may be achieved in two main ways: by wind or by insect. Insect-pollinated flowers tend to be colourful, scented and placed in an obvious place, the corolla often playing some part in selection of the pollinating insects, not merely a passive attractant. The selection of a pollinating agent may be achieved by elongation of the corolla into a tube with nectar at the base (suitable only for insects with long probosces), e.g. columbine. Other colourful corollas may have a form of jaw that needs pushing open (demanding the efforts of a large and powerful bee or similar insect), e.g. antirrhinum.

Some of the more bizarre lengths to which plants go to attract animals are shown in the illustration. In delphiniums the colour may attract different animals to their flowers. This is valuable for the species as it increases the chance of pollination should one agent be absent. The bee orchid, *Ophrys*, exhibits incredible mimicry, the male bee pollinating the flowers when it attempts to mate with the flower, presumably mistaking it for a female bee!

Additional features of insect-pollinated flowers are the sturdy build of all reproductive parts, a defence against the animals' movements, and the small amount of pollen produced. The pollen grains are large and sticky – an aid to attachment to the insect's body.

Wind-pollinated flowers are usually placed clear of leaves and lack petals and sepals which would obstruct free flow of air. The pollen is fine and dry and therefore easily carried, some grains often

Blue delphiniums attract insects, whilst red delphiniums attract some birds. (*Inset*) Bee orchid showing mimicry.

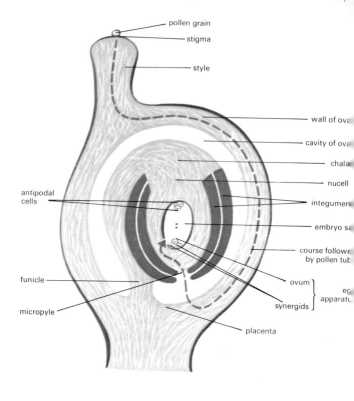

- pollen grain
- stigma
- style
- wall of ova[r]
- cavity of ova[r]
- chala[za]
- nucell[us]
- integumen[t]
- embryo sa[c]
- antipodal cells
- course followe[d] by pollen tub[e]
- funicle
- micropyle
- ovum
- synergids
- egg apparatu[s]
- placenta

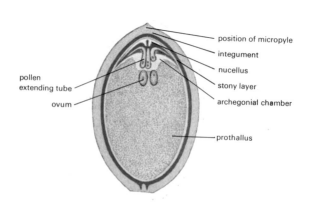

- position of micropyle
- integument
- nucellus
- stony layer
- archegonial chamber
- pollen extending tube
- ovum
- prothallus

increasing their surface area by sculptured exteriors, even having wing-like outgrowths.

Fertilization in plants

Although pollination may vary, the events following it are more uniform in nature. The pollen grain germinates, often as a result of the sticky, sugary secretion of the stigmatic surface. A thin pollen tube grows out from the pollen grain and bores its way down through the tissue of the style towards the ovule, gaining its nourishment from the stylar tissue. Just what controls or induces the growth to the ovule is not clear. At all events within this tube are three nuclei: a pollen tube nucleus and two male gamete nuclei. The male nuclei are the result of a nuclear division of a single nucleus and are therefore identical. When the pollen tube reaches the micropyle it ruptures, and about this time the pollen tube nucleus degenerates, permitting the two male nuclei to emerge.

The two male nuclei move to the edge of the embryo sac, and one of them enters, passing between the two synergid cells, and fuses with the ovum. This then is the act of fertilization in flowering plants. The second male nucleus also enters the embryo sac, usually via the 'gate' of the synergids, and fuses with the two polar nuclei, forming the primary endosperm nucleus. Thus three nuclei have taken part in the formation of the endosperm nucleus.

The endosperm nucleus divides many times, cell walls forming slowly, resulting in a loosely organized mass of endosperm tissue. This tissue develops at the expense of food materials in the nucellus, i.e. the bulk of the ovule. The endosperm in turn will provide food for the developing embryo that results from the zygote, i.e. the product of the act of fertilization.

In a few rare cases it has been shown that one of the antipodal cells may act as an ovum should the true gamete be nonfunctional for any reason. Also in a few cases, e.g. lilies, the polar nuclei are four in number, not two, as each polar nucleus originates from a cluster of four identical nuclei, and whereas there are usually only two of these clusters, forming the egg apparatus and the antipodals, in lilies these are duplicated. Thus there results four clusters, each of which contributes a polar nucleus. Following fusion with the second male nucleus, the primary endosperm nucleus of lily is therefore a composite of five fused nuclei.

(*Top*) Vertical section through an angiosperm carpel showing events following pollination, leading to fertilization. (*Bottom*) Vertical section through a gymnosperm ovule for comparison.

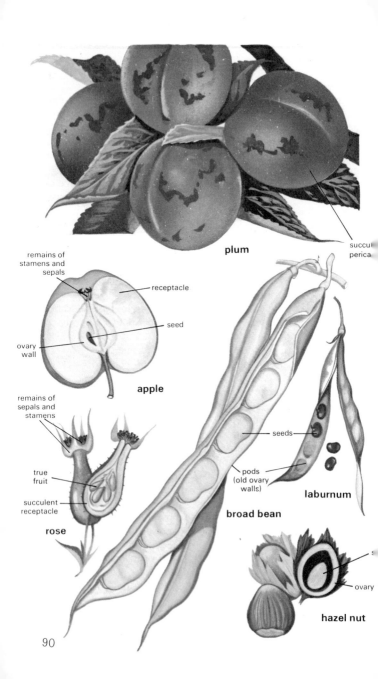

plum

succulent pericarp

remains of stamens and sepals

receptacle

seed

ovary wall

apple

remains of sepals and stamens

true fruit

succulent receptacle

rose

seeds

pods (old ovary walls)

laburnum

broad bean

ovary

hazel nut

90

Fruits and seeds

The developing embryo, mentioned on the previous page, gradually develops into a seed whilst the ovary from which it originated undergoes a typical series of changes. The main events during seed development are as follows. In the flower, the petals and sepals wither and drop off, their function complete, and the food they require can be put to better use. The stamens wither along with the stigma and style. In the ovary, the wall swells, as the seed grows, and changes in structure to form the pericarp or seed box. The pericarp may develop up to three different layers: the outermost epicarp, the mesocarp and innermost endocarp. This situation is seen very clearly in a succulent fruit such as the plum, where the epicarp is a coloured waxy skin, the mesocarp fleshy, and the endocarp a protective, hard stone. Often the pericarp remains dry, and as it matures, a considerable tension may develop within its walls – an aid to dispersal later, e.g. laburnum. Whatever the exact detail of the pericarp, it is always protective to some extent.

The seed

The seed is a complex structure partly derived from modified ovule tissue, and partly from new structures organized from the zygote's cell division.

The seed coat, or testa, develops from the hardened ovule skins, or integuments. The micropyle closes following fertilization and thus the testa is intact, although it will bear a scar (hilum) at the point of detachment from the ovule/seed stalk, the funicle.

The embryo soon becomes a structure with a number of different regions. The radicle, or future root, is one of the first features to become apparent, later followed by the smaller plumule, or future shoot. Endosperm accumulates within the developing embryo faster than it is used and is stored in structures known as cotyledons, such seeds being called non-endospermous. Dicotyledonous plants are those growing from a seed having two cotyledons, similarly a monocotyledonous plant originates from a seed with one cotyledon.

A few plants develop cotyledons but they do not store the endosperm within them, instead the embryo is encased in a greasy white mass of tissue, the endosperm. An example of such an endospermous form is the castor-oil plant where the cotyledons remain very thin and look remarkably leaf-like, even showing a branching system of veins.

Some typical fruits and seeds.

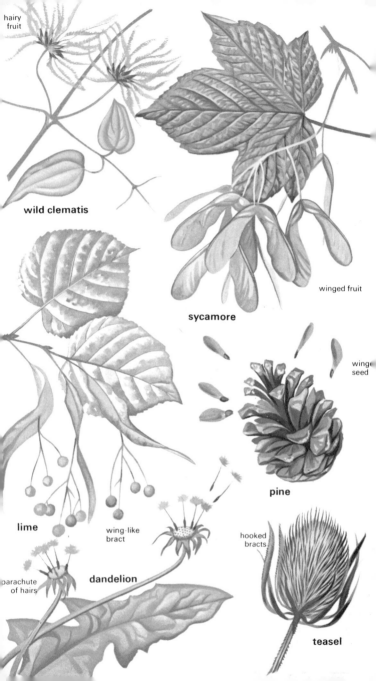

hairy fruit

wild clematis

winged fruit

sycamore

winged seed

pine

lime

wing-like bract

dandelion

parachute of hairs

hooked bracts

teasel

Fruit and seed dispersal

There are a number of reasons for dispersing fruits, seeds or spores, and they need not necessarily apply only to flowering plants.

A most obvious reason is to avoid **overcrowding** and this stresses the need for light for the leaves and soil for the roots. **Colonization** is a second important reason. It is a feature of a successful species of either plants or animals that they should be capable of reaching new territory, and then coping with its problems. The first step in colonization must be some effective dispersal mechanism.

There are four main agents of fruit and seed dispersal: wind, animal, water and mechanical action by the propagule itself.

Wind dispersal These fruits and seeds need to have a large surface area:volume ratio, or else need to be very small, in which case their surface area:mass ratio is large anyway. The means of increasing surface area are numerous, two types being hairs (parachute of hairs, e.g. dandelion; hairy seeds, e.g. willowherb), and wings (winged fruits, e.g. maple, ash; winged bract attached to fruit, e.g. lime; winged seed, e.g. pine).

Animal dispersal There are two main forms of dispersal by animals. Firstly, there are the forms that catch on to the outside of animals, due to hooks, spines or simply sticky exteriors to fruits and seeds. Secondly, there are the forms that are succulent and attract the animals as a food source, the propagule either passing through the gut of the animal unharmed, to be deposited in faeces at some distance from the parent plant, or the seeds prove too large to swallow whereupon they are spat out.

Hooked fruits include such types as burdock, where the entire inflorescence is detached due to hooked bracts; wood avens where a persistent style is hooked; and goosegrass where the pericarp develops hooks.

Succulent forms like plums rely on the resistant 'stone' (endocarp), with seeds within, being spat out. Tomato 'pips' are seeds with tough testas capable of passing through the gut.

Water dispersal This is an unusual method found in a few fruits. The coconut has a buoyant mesocarp, aided by a waterproof epicarp.

Mechanical dispersal This frequently relies upon tension set up in the pericarp to either spring two halves of the pericarp apart, e.g. pods of gorse, or to spring parts of a fused ovary apart, e.g. balsam. Poppy has a capsule formed by fused carpels with a slit-like opening which opens in dry weather and closes in wet conditions, thereby regulating the time of dispersal.

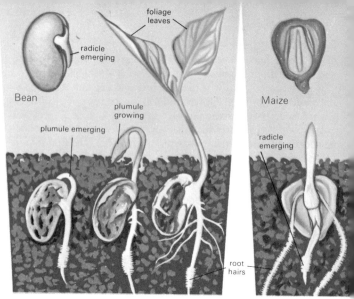

Bean
foliage leaves
radicle emerging
plumule growing
plumule emerging

Maize
radicle emerging
root hairs

Germination

The seed usually acts as a resistant stage in the life of the plant, and therefore will not start to grow (germinate) as soon as dispersed. Indeed, many seeds are capable of lying dormant for a few years until conditions are suitable for germination. The conditions necessary are firstly water, then the correct temperature and finally oxygen.

Water is required to soften the testa, to restore cell turgor, and to act as a medium for the cell chemistry. The correct range of temperature is important, firstly to ensure water is liquid and secondly to aid chemical reactions. The initial stages of germination will be visible as the swelling of the seed, due to uptake of water, but no apparent growth of organs will take place unless oxygen is present to enable aerobic respiration to occur, as germination is an energy-consuming process.

There are three common methods of germination, the differences being in the roles of the cotyledons.

Hypogeal germination

This is shown in the runner bean which is usually underground when germination starts, washed down by winter rains; however, this is not essential. The radicle emerges first, as in all cases, and quickly establishes a small root system. The plumule in its growth

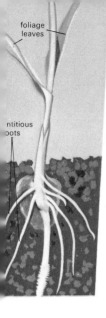

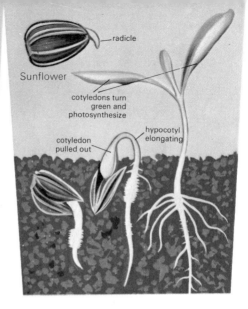

foliage
leaves

ntitious
oots

radicle

Sunflower

cotyledons turn
green and
photosynthesize

hypocotyl
elongating

cotyledon
pulled out

upwards will remain bent over as long as it is in the dark, presumably a protective growth form. Certainly beans germinating on the surface show a plumule that is green and is upright from its emergence. The first foliage leaf is a normal leaf, and the cotyledons (seed leaves) remain within the withering testa, their only job being to supply food for development.

Epigeal germination

The 'seed' of, for example, sunflower and sycamore is really the pericarp and seed within, but this need not concern us here. In epigeal germination the radicle emerges as usual, but at the base of the radicle there is a cylindrical region, the hypocotyl, which elongates pulling the cotyledons out from the testa (and pericarp in sunflower). Finally the cotyledons are dragged up to the light, where they fold back, turn green and photosynthesize, so that the cotyledons supply food and also photosynthesize.

Germination of cereals

The germination of this type is interesting as the solitary cotyledon remains underground as a food store, the radicle and plumule emerging and growing, protected in their early stages by sheaths, the coleorhiza and coleoptile, respectively.

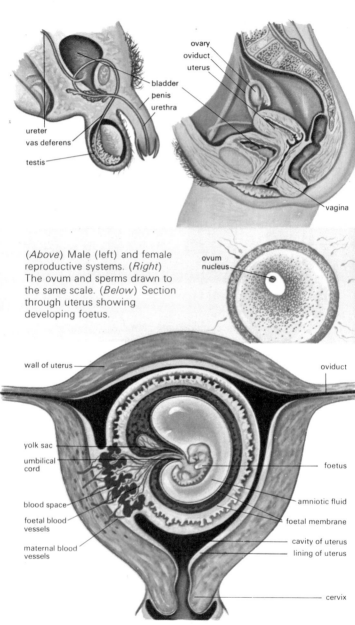

ovary
oviduct
uterus

bladder
penis
urethra

ureter
vas deferens

testis

vagina

(*Above*) Male (left) and female reproductive systems. (*Right*) The ovum and sperms drawn to the same scale. (*Below*) Section through uterus showing developing foetus.

ovum nucleus

wall of uterus

oviduct

yolk sac

umbilical cord

blood space

foetal blood vessels

maternal blood vessels

foetus

amniotic fluid

foetal membrane

cavity of uterus

lining of uterus

cervix

Sexual reproduction in man

Mention has already been made of the need for internal fertilization, following some form of protected transfer of the male gametes. The various forms of mating and copulation differ with the species concerned, but a closer look at sexual reproduction in man will give an idea of the solutions found in mammals generally.

In most, if not all mammals, there is some form of courtship or social behaviour pattern prior to mating. These behaviour patterns are often elaborate and help to form a reasonably stable association between male and female. In the case of man, the various ploys and moves encountered in courtship may take up a considerable time, even to the detriment of other human activities.

Female organs

The excretory and reproductive systems of females are separated far more than in males, as might be expected in view of the manner of development and birth. The ovaries produce eggs at regular intervals, usually alternating so that any one ovary will contribute an ovum every other cycle. The uterus wall prepares for the implantation of a zygote at every cycle, and if this is unforthcoming, will break down its swollen lining – the period of menstruation. The new menstrual cycle is taken as beginning from the start of menstruation, and varies in length from one woman to another. The length is readily influenced by the physical and psychological state of the woman but is approximately twenty-eight days, in which case the shedding of an egg (ovulation) occurs on the fourteenth day. The movement of the egg down the oviduct is by the beating of cilia lining the duct walls.

Male organs

The testes develop in sacs outside the body wall where it is cooler than the normal body temperature, since coolness is necessary for sperm maturation. The penis becomes stiff and swollen with blood, and in the act of copulation (intercourse) is placed within the vagina of the female. Rhythmic movements of the penis in the vagina culminate in the sperm being forced out by muscular contractions, depositing them at the top of the vagina and even within the uterus. From here, the sperm swim up the oviducts to meet the ovum, assisted by muscular contractions of the uterus and oviducts. Fertilization occurs when a sperm penetrates the cytoplasm of an ovum and fuses with the nucleus, forming a zygote. Fertilization only occurs in the upper region (fallopian tube) of the oviduct.

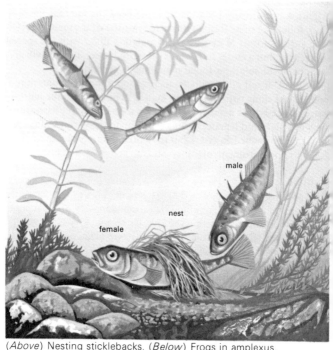

(*Above*) Nesting sticklebacks. (*Below*) Frogs in amplexus.

Development in man

The zygote forms a ball of cells which is passed down the oviduct and embeds in the uterus wall, this stage having taken seven days from fertilization. The embryo develops in a composite structure made up of three regions: the **foetus** (embryo); the **foetal membranes** which enclose the foetus in a fluid-filled chamber; and the **placenta**, a feeding device which obtains food and oxygen from the mother's blood system and, in return, passes waste materials into the maternal blood for excretion.

The foetus is attached to the placenta by an umbilical stalk containing blood vessels (see page 96). After approximately nine months gestation the baby is fully formed and is produced at *parturition* (birth). The muscular contractions of the uterus break the foetal membranes, the fluid within is expelled, followed later by the young child which is forced out via the cervix and vagina. The umbilical cord is cut and the baby must now feed and excrete for itself. A short time later the placenta, and remains of foetal membranes and umbilical cord, are expelled as the afterbirth.

The mother is normally capable of feeding the baby within a short time from birth, the breasts (mammary glands) producing milk containing a mixture of essential food materials.

In humans there follows a considerable number of years during which the child grows and matures, this period of life being a dependent stage upon the parents who thus exhibit considerable post-natal care and protection.

Sexual reproduction in other vertebrates

Aquatic vertebrates, like aquatic invertebrates, still frequently make use of external fertilization as there is no desiccation problem for the gametes. There is often some simple courtship, even in fish and amphibia, and in a few well documented cases, this may involve a complex ritual. Such a behaviour pattern is seen in the stickleback, in which the male lures the female to a nest that he has built, and encourages her to enter and lay eggs. Egg laying is induced by nudging the female's flanks. Following egg laying, the female is of no consequence and the male concerns himself with shedding sperm over the eggs.

Land colonists, such as the frog, which failed to evolve copulation must return to water to reproduce – a limiting feature in colonization. Here the male lures the female by croaking and induces egg laying by pressure from his forelimbs when he grips the female in amplexus.

blood vessels
carrying food from
yolk to chick

yolk sac

developing egg
tooth for breaking
out of egg shell

100

calcareous
shell

The frog is typical of the amphibia in its failure to overcome the gamete transfer problem, and this limitation to a region of water for reproduction is exacerbated by the amphibian skin respiration which requires a damp surrounding to prevent dehydration.

Reproduction in reptiles and birds

The reptiles evolved a number of features that have proved of great value, enabling them to colonize land far more successfully than amphibia. Firstly, they developed worthwhile lungs and an impermeable skin, a feature that freed their tie to damp regions. Secondly, the reproduction of reptiles involves internal fertilization with use of a penis in copulation. Furthermore, the female produces *shelled* eggs – an advance over most amphibia which lay naked eggs, another limitation to an aquatic development.

Birds display certain similar features to reptiles, namely copulation which ensures safe sperm transfer, and the female laying shelled eggs.

In both reptiles and birds the eggs are large and well supplied with yolk. They are coated with albumen soon after entering the oviduct and later enclosed in either a leathery (reptiles), or calcareous (birds) shell. These tough shells are lined by two egg membranes between which an air pocket is formed, at the blunter end of the egg.

After laying, the eggs develop provided the temperature is maintained at about 35°–38°C. In reptiles the heat is derived from the sun heating the soil or sand in which the eggs are buried. Occasionally the parents bury the eggs in a heap of rotting vegetation and rarely they will coil around them as an added protection, as in the python.

Courtship precedes mating and egg laying in reptiles, and parental care is provided mainly in the wealth of yolk supplied, although some females, such as those of turtles, do attempt to hide the eggs from predators. There is very little in the way of parental care after hatching and the slaughter of young, newly-hatched turtles by gulls is well known. Birds have a very elaborate courtship which may lead to a pairing for life, as in swans, and at the very least leads to a very strong bond for the immediate breeding season. The parents build a nest in which the eggs are incubated by the heat of the parent's body. After hatching there is a great deal of care provided by the parent birds which feed and protect the young.

(*Top*) Turtle eggs are laid at night and buried for protection.
(*Bottom*) Twenty-three day old turkey chick in egg.

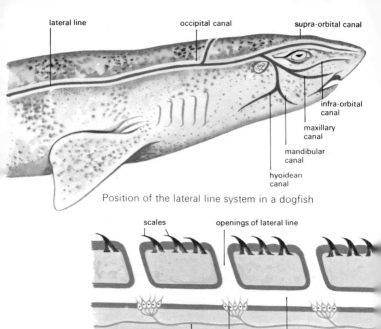

Position of the lateral line system in a dogfish

(*Above*) Vertical section through dogfish skin. (*Below*) Pressure waves radiating from point X reach A before B. The time difference enables fish to detect position of X.

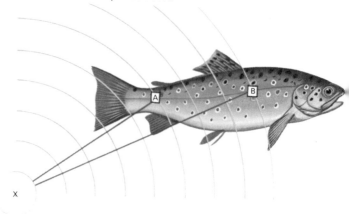

CO-ORDINATION IN ORGANISMS

As organisms became larger and more complex there was an increased need for methods of perception of changes both internally and from the outside world. In addition, an increase in overall size and complexity necessitated some internal communications network, in order that normal metabolism should continue without chemical activities interfering with some other process. Such an overall control would not only permit the larger organism to exist, but would ensure a more efficient way of life, as essential raw materials and products could then be utilized to the full.

In early multicellular forms the basic control was by diffusion of materials, i.e. a chemical communications network. Such a method is efficient for relatively small organisms, but when used by larger forms it becomes a rather slow process. Therefore, as animals evolved, an alternative communication system, the nervous system, also developed, and it can be seen that complexity of behaviour and awareness of the environment is parallelled by a well developed nervous system. The use of a chemical co-ordination system was retained and, as a transport system evolved, it was used to carry chemical messengers (*hormones*) from their site of formation to their place of activity.

One of the most primitive senses is that of being chemosensitive, i.e. aware of changes in the chemistry of the environment. The molecules of chemical compounds can range from being detected over the entire body surface, e.g. *Amoeba*, or in specialized regions of the body surface called olfactory organs – the lining of the nose. The olfactory senses of aquatic animals are very important as the chemicals are detected in the water film covering the surface of any olfactory organ. Many fish hunt using their olfactory organs, e.g. dogfish, and the anterior regions of their brains, the olfactory lobes, are very well developed. The use of scent as a means of hunting or detecting danger is also well developed in many land animals, e.g. dogs and deer, often being more important than eyesight in this respect.

Aquatic animals are very sensitive to localized changes in pressure in water, as any movement by prey or predator will cause such pressure waves. Hence fish and aquatic larvae of amphibia have a pressure-sensitive device, the lateral line, which enables them to pinpoint the source of disturbance. The system is thought to rely on the detection of pressure waves at more than one place on the body, and by a triangulation system, similar to radio-detector vans pinpointing transmitters, the source is determined.

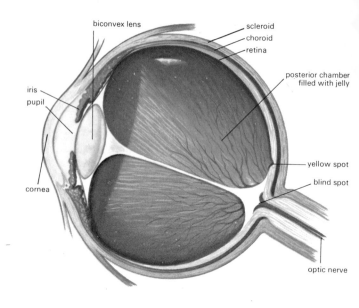

biconvex lens
scleroid
choroid
retina
posterior chamber filled with jelly
iris
pupil
yellow spot
blind spot
cornea
optic nerve

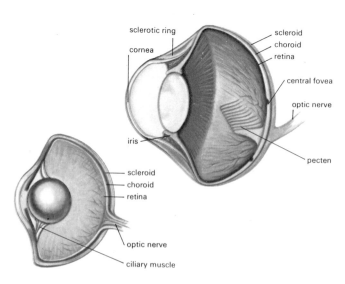

sclerotic ring
cornea
scleroid
choroid
retina
central fovea
optic nerve
iris
pecten
scleroid
choroid
retina
optic nerve
ciliary muscle

Eyes in vertebrates

The ability to detect light and to form images of the outside world is found throughout the animal kingdom. At the level of the protozoa, the entire body surface responds to wavelengths of light, sometimes with remarkable facility. For example, a small photosynthetic protistan, *Euglena*, can not only detect light, but will respond to different intensities, and can even exhibit a wavelength preference.

The possession of specialized light sensitive organs became necessary as animals progressed and became more complex. The form of these organs varies enormously from collections of light-sensitive cells, e.g. earthworm, to the complex compound eyes of arthropods. In the class of molluscs to which the squids belong, the most remarkable form of eyes in the invertebrates is found, its general plan being the same as the vertebrates in its possession of a biconvex lens, ciliary muscles, cornea, iris, etc.

The vertebrate eye is the finest optoreceptor in the animal kingdom, reaching the peak of its powers in the detection of moving objects by the birds, particularly the birds of prey.

The basic structure of some vertebrate eyes is shown in the diagram. The function of the different parts is as follows. The **lens** and **cornea** focus light rays on to the retina. The **retina** is responsible for the reception of light and, on being stimulated by light, sends an appropriate impulse to the brain. **Ciliary muscles** alter tension on suspensory ligaments and thereby change the shape of the lens, a means of altering focal length in mammals. The **iris** can vary the amount of light entering the eye by changing the size of the pupil.

The receptive retina occupies the same position in all vertebrates, although its powers differ. In some animals, e.g. monkey and man, the retina shows two sorts of cells – rods and cones. The presence of cones indicates the ability to see colours and not merely black, white and shades of grey. The cones aid in acuity of vision as they give an extra means of separating the constituents of an image. The disadvantage of cones, however, is that they only work in bright light and hence most birds, whose retina consists mainly of cones, are very nearly blind at night and hence must roost in a safe place until morning. Many nocturnal birds, such as owls, have a retina built almost entirely of rods which can function in low light levels, but gives only black and white vision.

(*Top*) The eyes of mammals focus by changing the shape of the lens. (*Bottom*) The eyes of fish focus by moving the lens. (*Centre*) The eyes of birds focus by changing the shape of the lens and the curvature of the cornea.

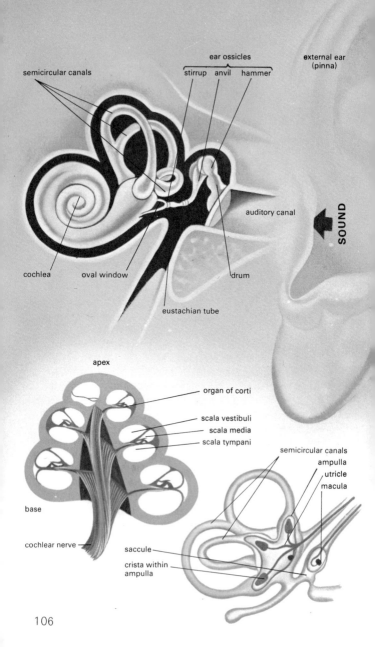

semicircular canals

ear ossicles
stirrup anvil hammer

external ear (pinna)

auditory canal

SOUND

cochlea

oval window

drum

eustachian tube

apex

organ of corti

scala vestibuli
scala media
scala tympani

semicircular canals
ampulla
utricle
macula

base

cochlear nerve

saccule

crista within ampulla

106

Ears

Aquatic animals use a variety of means to determine the source of movement in water. Terrestrial animals are also sensitive to pressure changes but in the air, these are converted into sound waves, i.e. they hear.

The ear varies from one class of vertebrate to another, the ability to 'hear' in fish being difficult to separate from their lateral line system (see page 102). Amphibia can detect sounds, for instance, the croaking of male frogs lures the female, but it is poorly developed. Reptiles differ in response to sound; lizards certainly 'hear', but snakes show very little, if any, response to airborne sounds, although they are very sensitive to vibrations that may be carried along the ground. Birds and mammals most certainly rely on hearing for a great deal of their normal behaviour pattern, for example the courtship and territorial song displays of birds, and the variety of sounds made by mammals either as warnings or as inducements. Mammals are unique in the extent to which their ears are developed; the external pinnae, or ear flaps, aiding location of sound as well as a funnelling effect on sound waves. The basic structure of a mammalian ear is illustrated.

The function of the parts of mammalian ears are as follows. The **pinna** is an aid to the listener in determining from which direction the sound waves come. The **middle ear** is a means of amplifying the sound waves which cause bulging of the ear drum. This causes movement in the three ear bones or ossicles. The **eustachian tube** equilibrates pressure within the ears with that of the atmosphere. The **semicircular canals**, by being in three planes, enable the fluid within (endolymph) to be affected by movement of the animal in any dimension. Any movement causes fluid to swirl through the canals and pass over sensory cells within the ampullae. The sensory hairs are held together by a gelatinous plate which swings, with the swirl of endolymph, like a swing door in a corridor. The **maculae** are sensitive patches of cells stimulated by the falling of calcareous particles, due to gravity, and hence inform the animal of its orientation. The **cochlea** is responsible for detecting different sounds and pitches of sound, i.e. this is the region of discerning hearing that is so sensitive in some animals. The cochlea consists of a coiled tube surrounded by, and set in, bone. Along the length of this tube run fibres, the tension and length of which are critical in the differentiation of different sounds.

(*Top*) Structure of the human ear. (*Bottom*) Detail of the cochlea (*left*) and the semicircular canals (*right*).

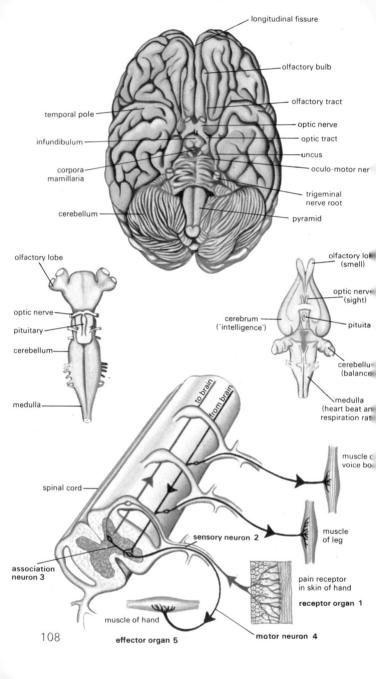

longitudinal fissure

olfactory bulb

olfactory tract

optic nerve

optic tract

uncus

oculo-motor ner[ve]

trigeminal nerve root

pyramid

temporal pole

infundibulum

corpora mamillaria

cerebellum

olfactory lobe

optic nerve

pituitary

cerebellum

medulla

olfactory lob[e] (smell)

optic nerve (sight)

pituita[ry]

cerebrum ('intelligence')

cerebellu[m] (balance)

medulla (heart beat an[d] respiration rat[e])

to brain

from brain

spinal cord

muscle [of] voice bo[x]

muscle of leg

sensory neuron 2

association neuron 3

pain receptor in skin of hand

receptor organ 1

muscle of hand

motor neuron 4

effector organ 5

108

Brains and the central nervous system

Nervous systems in animals range from the nerve nets of *Hydra* and the sea anemone, to the complex brain of a man. In such forms as *Hydra*, the nerve cells are capable of carrying impulses in any direction, and the spread of impulses is radial from the point of stimulus, i.e. any one stimulus is theoretically capable of directly stimulating all of the body, although response to it will normally be localized. The typical invertebrate nervous system, as seen very clearly in worms and arthropods, is a double, solid, ventral, ganglionated nerve cord joined to a pair of dorsal swellings (ganglia) in the region of the pharynx.

The vertebrate nerve cord is a single, dorsal, hollow tube swollen at its anterior end to form the brain. The brain plus the spinal cord form the core of the nervous system and constitute the central nervous system.

The basic nerve cell (*neuron*) transmits messages by means of minute changes in the conductivity of the cell membrane. Nerves responsible for carrying an impulse from the site of stimulus to the spinal cord are called sensory neurons (see diagram). Where a neuron joins another neuron there is a junction, or *synapse*. The synapse is useful as it acts as a valve ensuring one way flow of impulses; an impulse may 'jump across' in only one direction. The impulses must produce sufficient change in the synapse before they may pass, and hence a synapse is like an electrical resistance.

In a reflex arc there are five components: the receptor organ, e.g. skin; the sensory neuron; the connector or association neuron within the spinal cord; the motor neuron carrying the message to the final component; and the effector organ, e.g. muscle. Reflex actions permit many routine body actions to proceed without direct intervention by the brain. Such actions are coughing, sneezing, starting at loud noises, pulling hands away from hot objects, etc. Reflex actions are innate, i.e. they do not have to be learnt, they always elicit the same responses for any given stimulus, and finally, they are involuntary. If forewarned, however, a person can overrule the normal reflex response by control from the brain.

The brain regions have similar basic roles in all vertebrates, the main difference being in degree of development of the region. The functions of the regions are illustrated, the fish and mammal brains reflecting the range within the vertebrates.

Ventral views of human brain (top), dogfish brain (centre left) and rabbit brain (centre right). (*Bottom*) Reflex arc components.

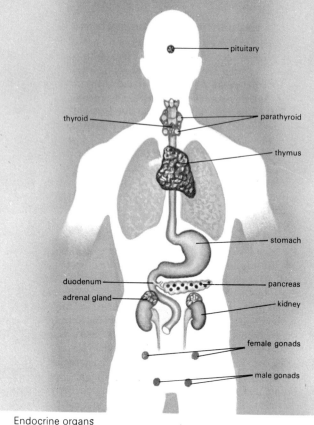

Endocrine organs

Chemical co-ordination in animals

There are a number of features concerning the use of hormones that should be kept in mind when thinking generally of hormonal and neural communication. Firstly, hormone action is far slower than nervous responses, and because hormones are carried freely in the blood, they can potentially affect any part of the body, whereas neural messages are carried by specialized cells (neurons) to specified regions. This last point explains why one hormone may be capable of widespread response in several organs at any one time; the nerve message may be very localized, even to a single muscle. Hormonal messages may be initiated and transmitted over long periods of time and hence can be used for very long term processes, e.g. growth.

Hormones are produced by endocrine organs, often called ductless glands, as their hormonal products are passed directly into the blood stream and not out via some duct. Most work on hormones has been done on humans and what follows will refer to man, although this appears to be true for mammals generally.

Thyroid gland This secretes thyroxine, a hormone that regulates the rate of metabolism and hence growth. Underactivity means a slow metabolism with a retardation of mental and physical development if it occurs in young children. Adults deficient in thyroxine become lethargic and exhibit swollen necks. Excess leads to a highly excitable and nervous state.

Adrenal glands A two-part organ, the outer cortex controls mineral balance in the body, the inner medulla produces the hormone adrenalin which causes marked changes in the body, for a short time. Adrenalin effects include increased blood pressure, increased respiration rate and heart beat, inhibition of digestive activity ('butterflies in stomach' effect), rise in blood sugar level, enlarged pupils.

Pancreas Part of the pancreas are the scattered clumps of cells known as the islets of Langerhans which produce insulin. Insulin causes a drop in blood sugar by enhancing its conversion to glycogen in the liver and tissues.

Pituitary organ This is the 'master' endocrine organ as it can stimulate most other endocrine organs. Among its effects are the following: stimulates the thyroid gland; ensures the adrenal cortex functions properly; can cause blood pressure and blood sugar to be raised, without the other side effects that adrenalin causes when it raises these levels; regulates growth, any upset causing either gigantism or dwarfing of the individual; stimulates the gonads to produce gametes and initiate their own production of sex hormones; induces milk production in nursing mothers and the production of ADH.

Gonads These produce sex hormones which regulate menstrual cycles in females and cause secondary sexual characteristics to develop, e.g. beards in men.

Thymus This is thought to play a part, indirectly, in antibody production by the lymphatic system, by the production of a stimulating hormone.

Pineal organ An organ of debatable function. It is thought to have some gonad regulating role.

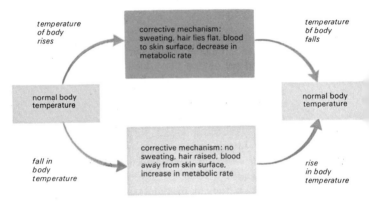

Flow diagram to show the homeostatic control of body temperature.

Homeostasis

Normal cell metabolism can only occur over a relatively narrow range of conditions. This range is probably determined by the conditions that prevailed when life evolved, and if, as seems likely, life evolved in the sea, then the chemical composition of the sea is probably the norm for life. The high specific heat of water meant that life evolved at a relatively stable temperature, whilst the sheer volume of the ocean meant that relatively little ionic change occurred due to dilution or evaporation.

Since life, and particularly in this context animal life, evolved on to land the environment changed dramatically. Not only did the physical problem of support and locomotion have to be overcome, but the physiological problems of changing temperature, evaporation and excretion now had to be solved. It is a reflection of how well they coped when we look to the most successful groups of vertebrates, the mammals and the birds. These two classes are both capable of maintaining a constant body temperature irrespective of environmental changes, i.e. they are homoiothermic. Other organisms are directly influenced chemically by any environmental temperature changes (poikilothermic), and hence only homoiotherms can live in the polar regions, as poikilotherms would have their cell chemistry inhibited by the extreme cold.

The example cited above shows the value of maintaining a constant internal state, namely it bestows greater ability to colonize new territory and hence reduces overcrowding whilst spreading the species. This ability to maintain a constant internal equilibrium is called *homeostasis* and covers not only temperature, but also such features as water balance, salt levels, nitrogenous waste control, carbon dioxide : oxygen ratio, etc.

In the limited space available it is only possible to mention one factor, namely temperature, and its regulation in humans.

Excess heat A person gets too hot for a number of reasons, but the responses are immediate. Firstly, heat is lost by increasing the radiant heat loss from the body and this is achieved in a number of ways. The hairs flatten, reducing the insulating layer of air and blood capillaries dilate bringing warm blood closer to the surface. Sweat is produced which cools the body when it evaporates. Internal changes, such as slowing the metabolic rate, aid in slowing down heat generation.

Decrease in heat This causes the reversal of the above features. The thermostat of the body is the hypothalamus, a ventral region of the brain adjacent to the pituitary, which detects changes in blood temperature as it flows through.

The illustration below shows how a dog brings its body temperature back to the normal after a rise in temperature due to activity.

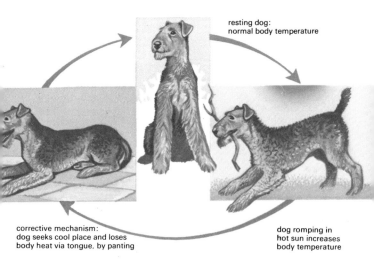

resting dog:
normal body temperature

corrective mechanism:
dog seeks cool place and loses
body heat via tongue, by panting

dog romping in
hot sun increases
body temperature

Dwarfism and gigantism is caused by a pituitary hormone imbalance.

Height of peas in cms

Dosage of gibberellic acid in µg (micrograms)

Homeostasis and hormones

The brief introduction to hormones on the previous page made mention of the manner in which the body communicates and achieves this steady state. In many cases hormones are used as the achievement of an equilibrium may well be a long process gained by a form of see-saw effect, i.e. correction leads to overcorrection, followed by a small tilt the other way until the particular factor is at the right level.

An example of this is the blood sugar balance and the use of hormones to hold it steady. The blood normally has 100 mg of glucose per 100 cm^3 of blood. Following a heavy meal rich in carbohydrates, the blood will contain higher levels of glucose than usual, particularly in the hepatic portal vein leading from intestine to liver. The presence of high glucose levels itself acts as a stimulant to the islets of Langerhans which produce insulin. Insulin is carried in the blood to the liver where it prevents the formation of further glucose from glycogen, and facilitates the conversion of glucose to glycogen. In addition, insulin increases the liver's rate of glucose breakdown and facilitates the storage of excess sugar as fat. These chemical activities quickly lower the glucose concentration in blood and as the level drops, so less insulin is produced. Eventually the required concentration is reached and insulin production drops to the minimum required to maintain the status quo, i.e. this shutting down of insulin production is an example of negative feedback.

The converse of the above is true, i.e. in time of glucose shortage the pancreas produces very little insulin and there is a conversion of glycogen to glucose. This last reaction can also be achieved by a pituitary hormone or, in times of stress or danger, by adrenalin.

Growth in plants and animals

The illustration of a dwarf and a giant shows what may happen if the pituitary organ is defective and produces either too little or too much, respectively, of its growth hormone. The growth is particularly noticeable in the bones.

Plants are stimulated differently; indol acetic acid enables the cells to swell by making their walls more plastic, i.e. reduces wall pressure and hence permits osmotic turgor pressure to inflate cells more. Alternatively gibberellic acid causes great elongation of the internodes by both cell elongation and some cell division as well, the quantity and concentration being critical.

The effects of gibberellic acid on the growth of peas.

ECOLOGY

Ecology is the study of plants and animals in their natural surroundings, their interaction with each other, and with the physical environment in which they live. The site where an organism lives is its habitat, and the relationship that exists between an organism and its habitat is very complex. The features helping to shape any habitat may be divided arbitrarily into three categories.

Climatic The physical climate will obviously have a big effect on any habitat and any organisms that attempt to colonize it. The two extremes of freezing polar cold and fierce equatorial heat must clearly influence any area of land or water.

Edaphic These are features associated with the soil, and here again the picture can be confused if the subsoil or bedrock is covered with a layer of material from another region, e.g. by glaciation. Nevertheless, soil types with their different degrees of acidity, texture, aeration, etc., will influence plant life and therefore, indirectly, animal life.

Biotic This is the effect that living organisms can have on an area. Such forms as fungi and bacteria can greatly influence the soil in their rate of breakdown of dead organic material. Grazing herbivores may change the cover in time, and this in turn will influence

Succession of plants in a community.

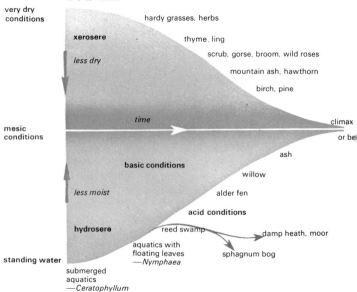

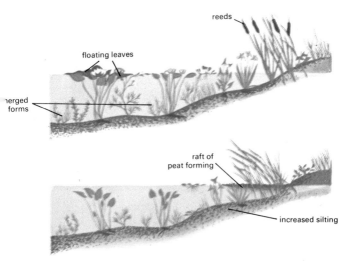

reeds

floating leaves

merged
forms

raft of
peat forming

increased silting

Two stages in the silting up of a pond, showing the advance of
plants from the margins.

the degree of erosion of top soil. Such simple examples give some
idea of the complex interaction between the three types of factors
mentioned above.

Succession

As organisms colonize an area, they alter it in many ways, for
example, the early colonizers die and their remains add to the
humic content of the site. These changes may be very slow, as for
example in the colonization of rocks by lichens, but eventually the
nature of the habitat will change. As it changes, an area may well
become more suitable for another type of colonizer, i.e. the original
'pioneer' forms have been succeeded by the next generation of
colonizing organisms. This displacement of one form by another
type is called *succession*, and will continue until the form living at
any one time alters the environment very little, or they themselves
are a self-perpetuating system. Such final colonizers are termed
climax types.

The climax population will be greatly influenced by such features
as the climatic (climatic climax), or the physical (edaphic climax)
nature of the soil. For any given climate there tends to be a given
climax vegetation and its attendant fauna, the original pioneer site
making little difference to the final outcome.

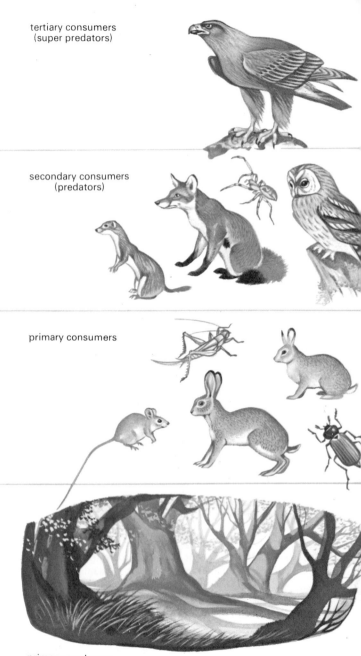

tertiary consumers
(super predators)

secondary consumers
(predators)

primary consumers

primary producers

Energy flow

A community is made up of many different species in varying numbers, and the study of communities can be undertaken at many different levels. Clearly the study of a large lake community with its thousands of species of animals and plants will take a great deal of time and manpower. This was the case in the Odell Lake study, carried out by the Oregon Game Commission which started in 1950 and ended in the early 1960s. Other communities studied have been overgrown orchards, town parks, restricted intertidal shore pools, etc.

Whatever the community, each species within it has its own place to live and role to perform. The place and function of a species is known as its *niche*, for example, freshwater crustaceans are scavengers and their niche is to roam the surface of mud and plants removing the decaying organic remains.

Any study of a community soon shows that its different members are linked, not simply by geographical constraints, but by an interdependent food web. That is, each species within a community depends on some other species for its food, and this feature binds them together.

The primary source of energy, upon which all life depends, is the sun, the form of energy being the sunlight. This solar energy is converted into chemical energy by green plants – the primary producers. The process of photosynthesis is the means of converting the light energy to chemical energy. The primary producers provide food for a wide range of herbivores which are called primary consumers. These primary consumers are themselves preyed upon by secondary consumers (carnivores). There may even be tertiary consumers which are the large carnivores, such as wolves. This description is merely a simplified, somewhat sterile indication of what is, in fact, an extremely complex and dynamic system. It is perfectly possible for a species to be found at more than one level in the account above, for example, bears will eat fruit (primary consumers), but it is also not unknown for them to eat fish, i.e. they are carnivores at the secondary or even tertiary level.

A further point to be stressed is the idea of a food 'web' existing, and not merely a series of linear food 'chains'. Each food chain is dependant on, and linked to other food chains in nature to form a complex net or web.

The pyramid of organisms in a food web.

A

C

D

E

120

Food webs

In the example quoted of a food chain, there would be several side branches and linking units. The primary consumers would be numerous and any one secondary consumer would rely upon several different species for its food. It is also possible that the secondary consumer (an insectivorous bird) would be preyed upon by numerous other carnivores, including rats which would eat its eggs. A very important point, so far omitted, is the role of the omnivorous scavengers living upon the dead representatives of all levels of the food web.

In a balanced, thriving community, the breaking of one link in a food web would not be too disastrous as other food sources would be available. It is only when some major event occurs, e.g. forest clearance or land drainage, that real problems arise, as very many threads and cross links in the web are broken.

A community disrupted by some natural disaster, for example, forest fire following lightning, or flooding of land, will establish new and tenuous food webs as soon as recolonization occurs.

Obviously, as succession progresses the changing flora and fauna mean that the interrelationships seen at any one time are purely short lived. Thus as succession and colonization change a habitat, so will the way in which energy flows through a community.

Adaptation

On page 117, the concept of succession was introduced and its dynamic nature has again been emphasized in the brief mention of food webs, where changing species result in a variation in mode of energy flow. The invading species of organisms will only succeed if their particular form and physiology is capable of coping with the prevailing conditions. Biologists frequently use the expression that an organism is 'adapted to its environment', and it is this potential to adapt, or to survive due to some feature in an organism, that makes it a successful colonist. The adaptive features of those dominant species in a climatic climax (see page 117) are often those which ensure self-perpetuation without changing the situation so dramatically as to enable another species to invade.

Even so, the status quo at any one time can change leading to spectacular population changes, e.g. the extinction of giant reptiles 100 million years ago.

Adaptations in climbing plants. A – ivy using roots; B – grape using tendrils; C – virginia creeper using sucker-like pads; D – bramble using thorns; E – convolvulus using a twining stem.

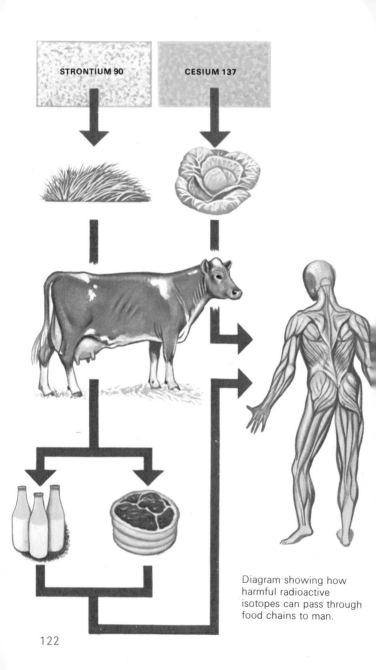

STRONTIUM 90

CESIUM 137

Diagram showing how harmful radioactive isotopes can pass through food chains to man.

122

There is insufficient room to give many examples of adaptive features found in successful species, but often a successful feature will be found in many species, for instance, the climbing habit of some plants enables more energy to be devoted to reproduction.

Man in nature

For successful survival an animal must be in harmony with its environment, and this is achieved by adaptation to any changes, whether natural or induced. Failure to adapt leads to extinction, no matter how long the previous harmonious period of success.

Man is a relatively recent arrival on the scene and so far his powers of adaptation have been phenomenal. Until some 10,000 years ago man was another hunter fitting in with other creatures in a complex food web. Then a change in his way of life appeared with the advent of agriculture and the domestication of livestock. This led to clearance of land and the introduction of small patches of monoculture, i.e. an area of a single species, his crop. In creating this civilized way of life, man started to deplete the land of its nutrients, and to interfere with the regular cycling of materials, and the small scale pollution of the rivers. This interference was relatively unimportant in the global picture as man's numbers were small. As time passed, however, the number of people increased, and their influence became increasingly detrimental to the ecosystem.

The major interference really started with the scientific revolution of the seventeenth century, and the traumatic effects of the last 150 years, the industrial age. The last few hundred years, and especially the last half century, have seen exploitation of the earth's natural energy resources on an unprecedented scale. Man has learned to use, and now rely upon, timber, coal, oil and natural gases. Not only has he removed them from their sites and transported them across the globe, but their use has polluted the other natural resources that he relies upon. One such resource is water, without which man will perish.

On page 112, the topic of homeostasis was discussed, and it was pointed out that man, being a mammal, is capable of controlling his internal environment. He is, however, unique in his ability to also control his external environment; how else could he exist in airless, frozen space or the freezing South Pole? It is this ability to modify the habitat that is both man's strength and his weakness. The values are obvious, and at last, perhaps, so are the dangers. In the last twenty years the problems of conservation and pollution have loomed large and, for successful survival, must be solved.

BOOKS TO READ

The publication of biological literature is prolific, but a few common and useful texts, to explain points raised in this brief introduction, are listed below.

General textbooks

Life: Form and Function by C. V. Brewer and C. D. Burrow. Macmillan, London, 1972.

Introduction to Biology by D. G. Mackean. John Murray, London, latest edition 1973.

General reading

The Plant Kingdom by Ian Tribe.

The Animal Kingdom by Sali Money.

Evolution of Life by Catherine Jarman.

The above three titles can be found in the Hamlyn all-colour paperback series.

For those wishing to read good general books restricted to a narrower range of material, the New Naturalist series, published by Collins, has a range of over fifty titles covering a wealth of biological topics.

INDEX

Page numbers in **bold** type refer to illustrations

TITLES IN THIS SERIES

Arts
Art Nouveau for Collectors/Collecting and Looking After
Antiques/Collecting Inexpensive Antiques/Silver for Collectors/
Toys and Dolls for Collectors

Domestic Animals and Pets
Cats/Dog Care/Dogs/Horses and Ponies/Tropical Freshwater
Aquaria/Tropical Marine Aquaria

Gardening
Flower Arranging/Garden Flowers/Garden Shrubs/House Plants

General Information
Aircraft/Beachcombing and Beachcraft/Espionage/Freshwater
Fishing/Modern Combat Aircraft/Modern First Aid/Photography/
Sailing/Sea Fishing/Trains/Wargames

History and Mythology
Witchcraft and Black Magic

Natural History
Bird Behaviour/Birds of Prey/Birdwatching/Butterflies/Fishes of the
World/Fossils and Fossil Collecting/A Guide to the Seashore/
Prehistoric Animals/Seabirds/Seashells/Trees of the World

Popular Science
Astrology/Astronomy/Biology/Computers at Work/Ecology/
Economics/Electricity/Electronics/Exploring the Planets/Geology/
The Human Body/Microscopes and Microscopic Life/Psychology/
Rocks, Minerals and Crystals/The Weather Guide